Test Pattern Generation using Boolean Proof Engines

Rolf Drechsler · Stephan Eggersglüß
Görschwin Fey · Daniel Tille

Test Pattern Generation using Boolean Proof Engines

Springer

Rolf Drechsler
Universität Bremen
AG Rechnerarchitektur
Bibliothekstr. 1
28359 Bremen
Germany
drechsle@informatik.uni-bremen.de

Görschwin Fey
Universität Bremen
AG Rechnerarchitektur
Bibliothekstr. 1
28359 Bremen
Germany
fey@informatik.uni-bremen.de

Stephan Eggersglüß
Universität Bremen
AG Rechnerarchitektur
Bibliothekstr. 1
28359 Bremen
Germany
segg@informatik.uni-bremen.de

Daniel Tille
Universität Bremen
AG Rechnerarchitektur
Bibliothekstr. 1
28359 Bremen
Germany
tille@informatik.uni-bremen.de

ISBN 978-90-481-8491-0 e-ISBN 978-90-481-2360-5
DOI 10.1007/978-90-481-2360-5
Springer Dordrecht Heidelberg London New York

Printed on acid-free paper

Springer is part of Springer Science+Business Media (www.springer.com)

Preface

After producing a chip, the functional correctness of the integrated circuit has to be checked. Otherwise, products with malfunctions would be delivered to customers, which is not acceptable for any company. During this post-production test, input stimuli are applied and the correctness of the output response is monitored. These input stimuli are called test patterns. Many algorithms for *Automatic Test Pattern Generation* (ATPG) have been proposed in the last 30 years. However, due to the ever increasing design complexity, new techniques have to be developed that can cope with today's circuits.

Classical approaches are based on backtracking over the circuit structure. They have been continuously improved by using dedicated data structures and adding more sophisticated techniques like simplification and learning. Approaches based on *Boolean Satisfiability* (SAT) have been proposed since the early 1980s. Comparisons to other "classical" approaches based on FAN, PODEM and the D-algorithm have shown the robustness and effectiveness of SAT-based techniques.

Recently, there is a renewed interest in SAT, and many improvements to proof engines have been proposed. SAT solvers make use of learning and implication procedures. These new proof techniques led to breakthroughs in several applications, like formal hardware verification.

In this book, we give an introduction to ATPG. The basic concept and classical ATPG algorithms are reviewed. Then, the formulation of this problem as a SAT problem is considered. Modern SAT solvers are explained and the transformation of ATPG to SAT is discussed. Advanced techniques for SAT-based ATPG are introduced and evaluated in the context of an industrial environment. The chapters of the book cover efficient instance generation, encoding of multiple-valued logic, use of various fault models and detailed experiments on multi-million gate designs. The book describes the state-of-the-art in the field, highlights research aspects and shows directions for future work.

Bremen, January 2009

Rolf Drechsler
drechsler@uni-bremen.de

Stephan Eggersglüß
segg@informatik.uni-bremen.de

Görschwin Fey
fey@informatik.uni-bremen.de

Daniel Tille
tille@informatik.uni-bremen.de

Acknowledgments

Parts of this research work were supported by the German Federal Ministry of Education and Research (BMBF) under the Project MAYA, contract number 01M3172B, by the German Research Foundation (DFG) under contract number DR 287/15-1 and by the Central Research Promotion (ZF) of the University of Bremen under contract number 03/107/05. The authors wish to thank these institutions for their support.

Furthermore, we would like to thank all members of the research group of Computer Architecture at the University of Bremen, Germany for their helpful assistance.

Various chapters are based on scientific papers that have been published at international conferences and in scientific journals. We would like to thank the co-authors of these papers, especially our collaborators Andreas Glowatz, Friedrich Hapke and Jürgen Schlöffel for their contributions and steady support. We would also like to acknowledge the work of Junhao Shi, who was one of the driving forces, when this project started and he was a PhD student in the group. We would also like to thank Arne Sticht, René Krenz-Bååth and Tim Warode for helpful discussions.

Our special thanks go to Michael Miller who spent a huge effort in carefully proof-reading and improving the final manuscript. Finally, we would like to thank Lisa Jungmann for helping with the layout of figures and for creating the cover design.

Contents

Chapter 1

Introduction

To make a long story short, finding tests to detect failures in complex circuits (*Automatic Test Pattern Generation*, ATPG) is a hard computational problem. Due to exponentially increasing circuit sizes, traditional ATPG algorithms reach their limits. Solvers for *Boolean Satisfiability* (SAT) recently showed their potential on real world ATPG problems. Thus, SAT-based ATPG is a promising alternative solution for this threatening bottleneck in circuit design when facing this rapid growth.

Why do we need to test? Why do current algorithms reach their limits? What is SAT? How may SAT-based ATPG help? The following longer story gives some answers.

Today almost every appliance we rely on is controlled by means of integrated circuits. This even holds for situations where our lives depend on the correct operation of such devices, e.g. when driving a car or at a traffic light. Failure of the control unit – the integrated circuit – can be a disaster and must be avoided. At the same time, the number of elements integrated in a circuit increases at an exponential rate according to Moore's law. This rapidly raises the computational difficulty of solving circuit design problems. Finding algorithms that guarantee the absence of failures in tomorrow's circuits with several millions of components is very demanding.

While designing a circuit, correctness is considered by means of powerful verification approaches. Typically, simulation-based techniques with large test benches and a huge investment in computational resources are necessary at the system level. Sophisticated techniques like constrained vector generation engines that increase the functional coverage to include special cases in the simulation are applied. At the lower levels, formal verification techniques are available to prove the correctness of a design with respect to predefined criteria. However, even if the circuit is designed correctly, physical faults introduced during production may affect the correct behavior of a device.

Besides an image loss when shipping faulty devices, a semiconductor company might also face financial losses if compensation has to be paid for faulty devices. Therefore, considerable effort is spent in post-production test to decide whether a device is free of faults or defective. Depending on the application area and the maturity of the production process, up to 30% of the costs of a device are attributed to post-production testing. A range of testing methods is necessary to filter out defective devices as well as those with reliability issues leading to an early failure in the field.

These methods typically involve the application of functional tests at speed and the use of dedicated test vectors to discover pre-defined types of faults. After each processing step, a device is tested to filter defective devices as early as possible, e.g. on the wafer or after packaging. This helps to reduce costs and to improve the location of weaknesses in the production flow. Functional at speed tests can often be derived from the test benches that are used for the verification of a circuit.

The generation of test vectors to detect pre-defined types of faults is usually called *Automatic Test Pattern Generation* (ATPG). For this purpose, a fault model and an abstract circuit model are typically used. An abstract circuit model is required to reduce the computational complexity. Essentially, any potential physical irregularity during production is of interest, since this might induce an irregular behavior during application. But obviously, there are too many potential physical irregularities to consider all of them and computing the resulting effects for a very accurate physical model is too computationally expensive. Instead, a gate level model is typically used. Mostly, Boolean reasoning is sufficient to model faults at this level and to generate test vectors. Moreover, this level of abstraction provides enough information about connectivity and functionality to model the effects of a large number of relevant physical faults.

A fault model is applied to the gate level description of the circuit. Different fault models are available to mimic different types of physical faults. The *Stuck-At Fault Model* (SAFM) is routinely applied. It models static functional faults such as the short of a wire to power or ground. Besides static faults, dynamic faults are also considered that do not modify the combinational function of a circuit, but the timing behavior. These so called delay faults model late arrival of transitions, i.e. signal changes arrive later than required by the clock frequency specified for the circuit.

In the domain of delay test generation, the application of the *Path Delay Fault Model* (PDFM) and the *Transition Delay Fault Model* (TDFM) is widespread. The PDFM captures large as well as small delay defects but the number of faults that have to be considered per circuit is typically very large. In contrast, the number of faults according to the TDFM is proportional to the number of elements in a circuit. However, if only the TDFM is applied, some dynamic faults may remain undetected. In practice, all of these fault models are combined to achieve a high coverage of physical defects.

Despite the abstraction achieved by considering fault models at the gate level, ATPG is computationally hard. Even for the relatively simple SAFM, deciding whether a test vector exists for a given stuck-at fault in a combinational circuit, is an NP-complete problem. Considering sequential circuits makes the problem even harder. Therefore, in practice the problem is simplified to a combinational problem. "Scan chains" are created by connecting all state elements in large shift registers. In test mode, values can be shifted into these registers. This "only" leaves the NP-complete combinational ATPG problem that has to be solved for all modeled faults for circuits with several million gates.

To cope with this computational complexity, sophisticated ATPG frameworks have been developed. The ATPG framework creates test vectors for all faults that can be detected. Those faults that do not have a test vector, i.e. cannot be detected, are proven to be untestable. Untestability may occur due to the structure of a circuit.

The main engines of these frameworks are *random pattern generation*, *fault simulation* and *deterministic pattern generation*. Figure 1.1 shows a simple ATPG framework. Random pattern generation followed by fault

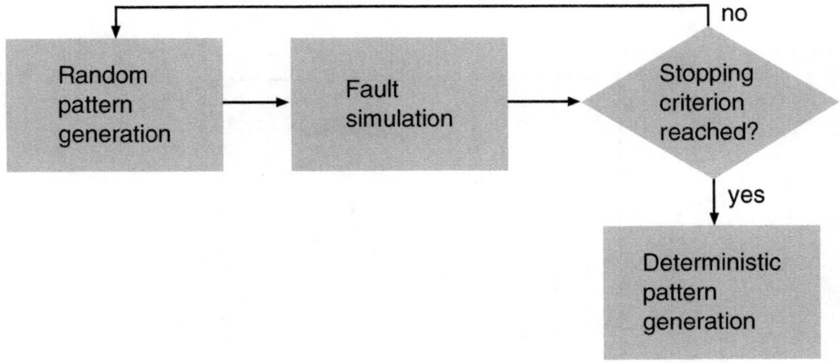

Figure 1.1: A simple ATPG framework

simulation runs until a stopping criterion is reached, e.g. no additional faults can be detected. Then, deterministic pattern generation classifies the remaining faults as testable or untestable. Practical ATPG frameworks have a much more sophisticated architecture, e.g. to insert scan-chains or to generate a small set of test vectors.

Random pattern generation is very efficient. Here, random input stimuli are generated for the circuit. Typically, a large portion of the faults is detected easily for any practical circuit. *Fault simulation* then determines the faults detected by a given test pattern. For this purpose, the input assignment is simulated on models of the correct circuit and the faulty circuit. Whenever simulation shows a discrepancy between the correct and the faulty circuit, the fault is detected by the given input assignment.

Of course, more sophisticated algorithms are available. Nonetheless, even for this simple approach the computational complexity is linear in the number of elements of the circuit. These two engines help to classify a large portion of faults as testable very easily. But untestability cannot be proven using a simulation engine and, in addition, the excitation of faults where only a few test vectors are available is almost impossible. In these cases, a deterministic engine is required.

Deterministic pattern generation corresponds to the NP-complete decision problem mentioned above. Thus, algorithms for deterministic pattern generation cannot be expected to run in polynomial time. A range of quite effective algorithms has been proposed. Typically, these algorithms exploit the structure of the problem. The task is to find a test vector for a given fault in a circuit described at the Boolean level. Figure 1.2 illustrates the search problem. Some immediate observations help to describe an algorithm.

The fault must be excited to become visible, i.e. the logic values in the correct circuit and the faulty version at the site of the fault must be different.

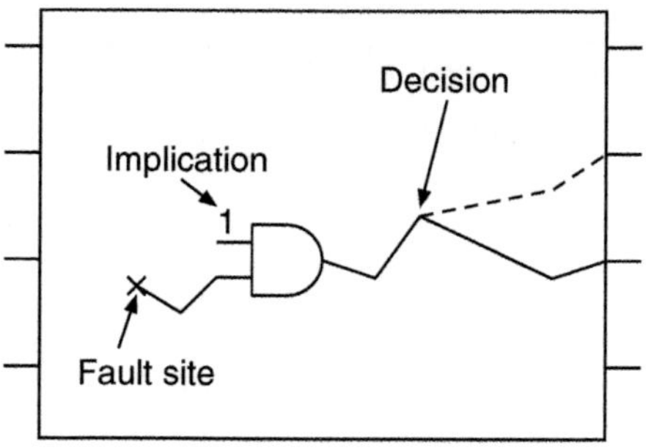

Figure 1.2: Justification and propagation

Moreover, in practice, faulty operations can only be observed at the outputs. Formulating such conditions at the Boolean gate level is straightforward. For example, passing a fault observation along one input of an AND-gate towards the outputs requires the other input to have the value 1 – otherwise the output of the AND-gate takes the value 0. Thus, a value is implied in the current state of the search as illustrated in Figure 1.2. Similar conditions can be formulated for all other types of Boolean gates.

At some stages during this process, decisions may be necessary, e.g. which paths towards outputs are used to propagate the fault observation. Such a decision may be wrong and may lead to a "dead end", i.e. no detection of the fault is possible under the current value assignments in the circuit model. In this case, the decision has to be undone together with all implications resulting from this decision – this is called backtracking.

Thus, an algorithm for deterministic pattern generation traverses the search space until a test pattern is found. If no test pattern can be found, the fault is untestable. To make this algorithm robust and applicable to large circuits, intelligent heuristics are necessary to make "good decisions" and powerful implication engines are required to efficiently derive the consequences of a decision.

Based on the first algorithm that worked on the circuit structure in this manner – the D-algorithm – several improvements have been proposed. These improvements were concerned with better heuristics, more powerful implication engines and learning from "dead ends". The algorithms to classify dynamic faults are very similar, but the computational effort is even higher since two time frames have to be considered for a single dynamic fault, i.e. one time frame for the initialization and one time frame for the propagation. Today, these algorithms often still classify a large portion of faults in acceptable run times. However, due to the exponential growth of circuits, the limits have been reached. Already today, for some very large or very complex circuits, a large number of faults remain unclassified.

In practice, this means, either the fault is untestable – which may cause a reliability problem and when too many faults are untestable, a design change is demanded, or alternatively, a testable fault is not tested. During application of the circuit an undetected fault may cause a control unit to fail. Therefore, more effective engines for deterministic pattern generation are required. Using solvers for *Boolean Satisfiability* (SAT), as discussed in this book, is a promising solution that has already proved its power in practice.

Deciding the satisfiability of a Boolean formula is a well-known problem in theoretical computer science. The question is whether there exists an assignment for a Boolean formula such that the formula evaluates to 1. This problem was the first proven to be NP-complete by Stephen A. Cook in 1971. This "negative" result discouraged research on early algorithms to solve the problem. Very powerful algorithms were developed not until the 1990s – even though these algorithms are effectively a very intelligent extension to those proposed in the 1960s. Learning was one of the keys to tackling problem instances of large sizes. As a consequence of these improvements, SAT solvers have been applied very successfully to formal verification of circuits as well is in other problem domains.

The use of a SAT solver as a powerful black-box engine for Boolean reasoning as shown in Figure 1.3 is very appealing. The original problem has to be transformed into an instance of the SAT problem. Then, the SAT solver provides a satisfying assignment or proves that no such assignment exists. Finally, a reverse transformation is used to extract the solution from the satisfying assignment.

Provided that both transformation steps can be done efficiently, the capability of the SAT solver to solve the problem decides whether this approach is feasible or not. Of course, using the SAT solver as a black-box does not yield the maximum performance. Knowing the structure of the original problem can help to improve the transformation into a SAT instance. Moreover, the SAT solver uses efficient algorithms for decision making and learning. Enabling an exchange of information between the SAT solver and the surrounding framework is crucial to gain higher performance.

In the context of this book, deterministic pattern generation is the original problem and a test vector for a given fault is the solution. Starting from this simple formulation, optimizations are essential to allow for a successful application to industrial circuits in a commercial ATPG framework. Throughout this book, the SAT-based ATPG tool PASSAT, that has been developed over the past 4 years by the group of Computer Architecture at the University of Bremen, Germany, is discussed. PASSAT has been integrated as a prototype into the ATPG framework of NXP Semiconductors

Figure 1.3: SAT solver as a black-box

(formerly Philips Semiconductors), that has been developed over 25 years. In this framework, PASSAT has been proven to be a powerful engine orthogonal to classical ATPG algorithms. PASSAT increases robustness and efficiency of the overall framework.

The book is structured as follows:

- *Chapter 2 – Preliminaries* introduces basic notations to describe circuits, a number of fault models and revisits traditional ATPG algorithms. Moreover, the industrial benchmark circuits considered throughout this book are briefly introduced.
- *Chapter 3 – Boolean Satisfiability* considers the SAT problem. After defining the problem, the techniques in today's state-of-the-art SAT solvers are discussed. Finally, the transformation of a gate level circuit into SAT constraints is shown and potential pitfalls are discussed.
- *Chapter 4 – SAT-Based ATPG* explains the basic transformation of ATPG to SAT and introduces optimizations in terms of embedding structural information into the SAT instance.
- *Chapter 5 – Learning Techniques* lifts the conflict-based learning from the SAT solver to the level of the ATPG tool. Techniques for application level learning from subsequent similar SAT instances are discussed and applied to ATPG.
- *Chapter 6 – Multiple-Valued Logic* explains how to handle real world constraints in an industrial setting in the ATPG tool. Environment constraints and tri-state elements require a multiple-valued logic encoded in the Boolean SAT instance. Trade-offs between different encodings are evaluated in detail.
- *Chapter 7 – Improved Circuit-to-CNF Conversion* shows efficient ways to reduce the size of a SAT instance derived from a circuit. In the industrial setting, only a few elements require a four-valued model while most elements only have Boolean values. Moreover, most faults can be observed at multiple outputs, but one observation point is sufficient. Exploiting these properties during CNF generation significantly shrinks the CNF size and also decreases the run time.
- *Chapter 8 – Branching Strategies* proposes problem-specific decision heuristics for the SAT solver that are tuned for ATPG. A good trade-off between small run times and a small number of unclassified faults is achieved.

- *Chapter 9 – Integration in Industrial Flow* discusses the embedding of PASSAT into an industrial ATPG framework. Completely replacing the highly tuned engines and accompanying algorithms for fault simulation, test compaction etc. is not feasible. Instead, a good way of using PASSAT within the framework is presented.
- *Chapter 10 – Delay Faults* extends the SAT-based approach. The model has to reflect the circuit's behavior over (at least) two time frames. Additionally, a mechanism to prevent hazards or race conditions is required; otherwise the fault effect may be missed if the observation happens at the wrong moment. Multiple-valued logics are proposed that model the required behavior. A procedure to identify efficient Boolean encodings for these logics is further shown.
- *Chapter 11 – Summary and Outlook* recapitulates the techniques proposed in this book and presents uncovered questions and directions for further research in SAT-based ATPG.

Chapter 2

Preliminaries

This chapter provides the necessary background to introduce the ATPG problem. First, circuits are presented in Section 2.1. This includes a brief description of how a sequential ATPG problem is reduced to a combinational problem. Furthermore, static as well as dynamic fault models are introduced in Section 2.2. In Section 2.3, a brief overview of a simple ATPG framework and fault simulation is presented. Classical ATPG algorithms working on the circuit structure for stuck-at faults as well as for path delay faults are presented in Section 2.4. Lastly, in Section 2.5, a description of how experiments are conducted in this book is given. The presentation of this chapter is kept brief, for further background, we refer the reader to text books on testing like e.g. [14, 59].

2.1 Circuits

In the following, a circuit C is assumed to be composed of the set of basic gates shown in Figure 2.1. These are the well-known gates that correspond to Boolean operators: AND, OR, XOR and NOT, where each gate has at most one successor. Additionally, FANOUT gates model fanout points in the circuit. Thus, a FANOUT gate always has a single predecessor and multiple successors. FANOUT gates help to easily identify fanout points in the circuit and to model individual faults on fanout branches.

The connections between gates are defined by an underlying graph structure. Throughout this book, gates are denoted by lower case Latin letters $a, b, c, \ldots$. Where, for the sake of convenience, additional indices may be used. For example, the gates in a circuit, in a set or along a path may be denoted by $g_1, g_2, \ldots, g_o$. Gates denoted by $i_1, \ldots, i_n$ and by $o_1, \ldots, o_m$ refer

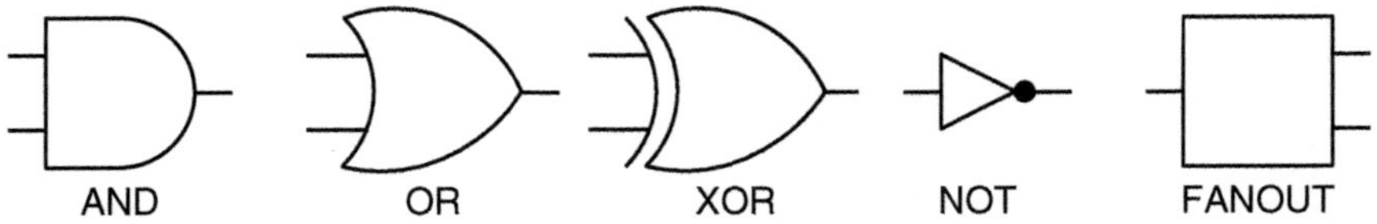

Figure 2.1: Basic gates

to primary inputs and primary outputs, respectively. A gate and the output signal of the gate are denoted by the same letter. Moreover, when a variable is associated to a gate or a signal, this letter is typically used as well.

The *transitive fanin* of a gate g is denoted by $\mathcal{F}(g)$. The transitive fanin of a set of gates $\mathbb{G}$ is denoted by $\mathcal{F}(\mathbb{G})$.

Extending the underlying library to other Boolean gates, if necessary, is straightforward. For gates that represent non-symmetric functions (e.g. multiplexers or tri-state elements), a unique order for the inputs is given by ordering the predecessors of a gate.

In most cases, only combinational circuits are considered. In these cases, all memory elements are split into a pseudo primary input and a pseudo primary output. Logically, these pseudo primary input and pseudo primary output can be considered as normal inputs and outputs, respectively. Where sequential circuits are considered, the required additional notation is introduced explicitly.

2.2 Fault Models

After producing a chip, the functional correctness of this chip with respect to the Boolean gate level specification has to be checked. Without this check, an erroneous chip would be delivered to customers which may result in a malfunction of the final product. This, of course, is not acceptable. On the other hand, a large range of malfunctions is possible due to defects in the material, process variations during production, etc. Directly checking for all possible physical defects is not feasible. Therefore, an abstraction in terms of a *fault model* is applied. In the following, the most common fault models are introduced.

2.2.1 Stuck-at Faults

The *Stuck-At Fault Model* (SAFM) [10] is well-known, well-understood and widely used in practice. The SAFM models static or permanent functional faults. In this fault model, a single line is assumed to be stuck at a fixed

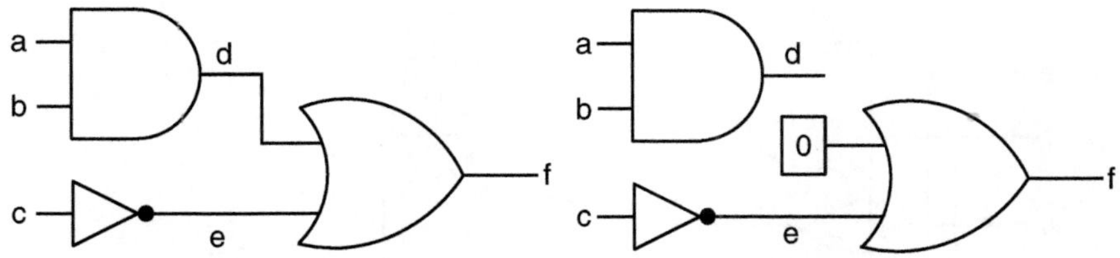

(a) Correct circuit (b) Faulty circuit

Figure 2.2: Example for the SAFM

value instead of depending on the input values. When a line is stuck at the value 0, this is called a *stuck-at-0 fault* (s-a-0). Analogously, if the line is stuck at the value 1, this is a *stuck-at-1 fault* (s-a-1). A stuck-at fault is denoted by a pair (g, val), where g denotes a signal (the output of a gate g) and val denotes the stuck-at value. For FANOUT gates, an additional index i identifies for which FANOUT branch the fault is modeled.

Example 1 *Consider the circuit shown in Figure 2.2a. When a stuck-at fault* $(d, 0)$*, i.e. the s-a-0 fault on line d, is introduced, the faulty circuit in Figure 2.2b results. The output of the AND gate is disconnected and the input of the OR-gate constantly assumes the value* 0*.*

ATPG is the task of calculating a set of test patterns for a given circuit with respect to a fault model. A test pattern for a particular fault is an assignment to the primary inputs of the circuit that leads to different output values depending on the presence of the fault. Calculating the Boolean difference of the faulty circuit and fault free circuit yields all test patterns for a particular fault. This construction is similar to a *miter* circuit [9] as it can be used for combinational equivalence checking.

Example 2 *Again, consider the s-a-0 fault in the circuit in Figure 2.2. The input assignment* $a = 1$, $b = 1$, $c = 1$ *leads to the output value* $f = 1$ *for the correct circuit and to the output value* $f = 0$ *if the fault is present. Therefore, this input assignment is a test pattern for the fault* $(d, 0)$*. The construction to calculate the Boolean difference of the fault free circuit and faulty circuit is shown in Figure 2.3.*

When a test pattern exists for a particular fault, this fault is classified as being *testable*. Otherwise, the fault is called *untestable*. The decision problem whether a fault is testable or not is NP-complete. The aim is to classify all faults and to create a set of test patterns that contains at least one test pattern for each testable fault.

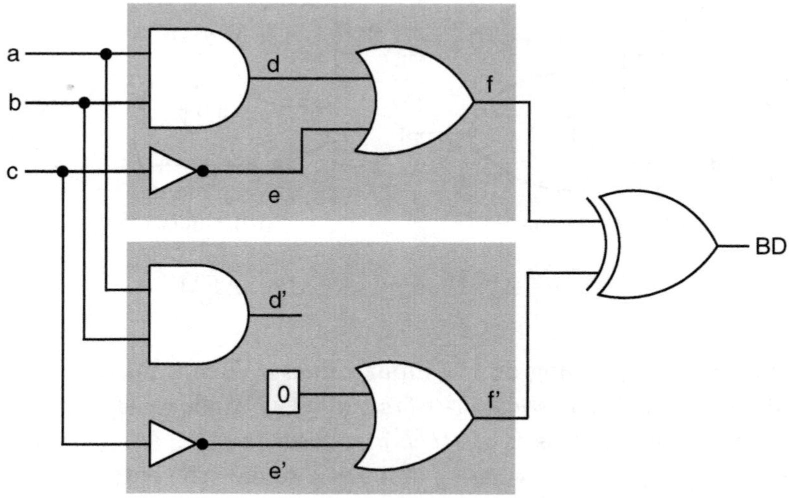

Figure 2.3: Boolean difference of faulty circuit and fault free circuit

Fault equivalence and *fault dominance* relations help to improve the performance of ATPG. Instead of considering all faults according to the given fault model, only a subset has to be enumerated. Two faults are said to be equivalent if they are tested by the same set of test patterns. Whenever one of the faults is detected, the other fault is detected as well.

Calculating this equivalence relation over the set of all faults is hard, but locally deciding fault equivalence can be done very efficiently. Consider an AND gate. The s-a-0 at the output is only detected by setting all inputs to 1. There are no other test patterns for this fault. Analogously, the s-a-0 at one of the inputs is detected by the same test pattern and no other test patterns exist. Therefore, the s-a-0 at the output and at any input are equivalent, and only one of these faults has to be considered.

Fault dominance is a similar concept. A fault A dominates another fault B if all tests for A also detect B. In this case, the dominated fault B does not have to be considered – it is detected whenever A is detected. Consider an AND gate with n inputs, the s-a-1 A at input i and the s-a-1 B at the output. Any test pattern with at least one 0 detects B, but only the test pattern with a 0 at input i and 1s at all other inputs detects A. Therefore, A dominates B.

The notions of fault equivalence and fault dominance can easily be extended to other types of gates. Typically, a fast preprocessing step called *fault collapsing* is used to remove all but one equivalent fault and all dominated faults according to some local structural analysis. The structural analysis described above is an example. This process is not complete, i.e. there might remain equivalent or dominated faults, but it is very fast. Moreover, the technique is quite effective, since most equivalence and dominance relations are detected.

Besides the SAFM, a number of other fault models have been proposed. The cellular fault model [39] models a malfunction within a cell consisting of one or more gates. By this, the function of the cell is changed. The bridging fault model [62] assumes an unwanted resistive connection between two gates. Here, a resistance of 0 Ω indicates a static fault – the two connected lines settle to the same logic value. These fault models mainly cover static physical defects, like opens or shorts. Dynamic effects are covered by delay fault models.

2.2.2 Delay Faults

The *Path Delay Fault Model* (PDFM) [94] models a distributed delay on a path from a (pseudo) primary input to a (pseudo) primary output of a circuit. When the cumulative delay of all gates and wires along the path exceeds the time for a clock cycle, a *Path Delay Fault* (PDF) occurs. The effect of a physical fault may be different for rising and falling transitions. Therefore, these two cases are modeled by the PDFM.

Formally, a PDF is given by $F = (P, T)$, where $P = (g_1, \ldots, g_n)$ is a path from a (pseudo) primary input g_1 to a (pseudo) primary output g_n. The type of transition is given by $T \in \{R, F\}$, where R denotes a *rising* transition and F denotes a *falling* transition. A rising transition goes from logic 0 in the *initial* time frame t_1 to logic 1 in the *final* time frame t_2, whereas the falling transition goes from logic 1 in t_1 to logic 0 in t_2.

To detect a fault, two test vectors v_1, v_2 are needed to propagate a transition along the path P during two consecutive time frames t_1, t_2. Note, that the transition must be inverted after an inverting gate on the path. The initial vector v_1 sets the initial value of the transition in time frame t_1, whereas the final vector v_2 launches the transition in t_2 at operating speed. For the case of a delay fault, the expected value cannot be observed at g_n. If multiple delay faults are present in the circuit, a test might not detect the fault because other delay faults may mask the targeted PDF.

The quality of a set of tests can be classified by the concept of robustness [18]. A test is called *robust* if, and only if, it detects the fault independently of other delay faults in the circuit. *Non-robust* tests guarantee the detection of a fault if there are no other delay faults in the circuit. If there is neither

Table 2.1: Off-path input constraints

	Rising rob.	Falling rob.	Non-rob.
AND/NAND	$X1$	$S1$	$X1$
OR/NOR	$S0$	$X0$	$X0$

a non-robust nor a robust test, the PDF is untestable. A detailed discussion of the classification of PDF tests and further sensitization criteria can be found in [63].

Robust and non-robust tests differ in the constraints on the off-path inputs of the path as shown in Table 2.1 (also known as sensitization criteria). An off-path input of P is an input of gate $g_i, i \in \{1, \ldots, n\}$ that is not on P. The values shown in Table 2.1 correspond to the seven-valued logic $L_7 = \{S0, S1, \overline{S}0, \overline{S}1, X0, X1, XX\}$ originally proposed in [71] and used for robust path delay test generation in [17].

For a non-robust test, it is sufficient that the off-path inputs have a *non-controlling value* (ncv) at t_2 (denoted by $X0/X1$). For robust tests, if the on-path transition on g_i goes from the non-controlling value to the *controlling value* (cv) of g_{i+1}, the off-path inputs of g_{i+1} must have a static non-controlling value (denoted by $S0/S1$). A robust test is also a non-robust test, but not vice versa. Applying static values to the off-path inputs avoids the situation that other delay faults on the inputs of the gate may have an influence on the value of the output.

Example 3 *Consider the circuit depicted in Figure 2.4 and the path $P = (a, d, e, g)$ with a rising transition. The off-path inputs of P are b, c, f. Under the non-robust sensitization criterion, the off-path inputs b, c have to be fixed to $X1$ because the non-controlling value of d and e is 1. Off-path input f has to be fixed to $X0$; the non-controlling value of g is 0. The corresponding input assignment denoting a non-robust test of the PDF $F = (P, R)$ is $a = R$, $b = X1, c = X1$.*

The PDFM is very powerful and accurate but typically a circuit contains too many paths to consider all PDFs – the number of paths may be exponential in the number of gates. Thus, in practice, only critical paths – those with the least slack – are considered with respect to the PDFM. Critical paths can be extracted by dedicated algorithms, see e.g. [56, 105].

Numerous relaxed delay models have been introduced. All of them are less accurate than the PDFM in terms of coverage of physical defects, but provide a better fault coverage in practice. Essentially, instead of considering paths, shorter segments are considered. Then, the fault effect is considered

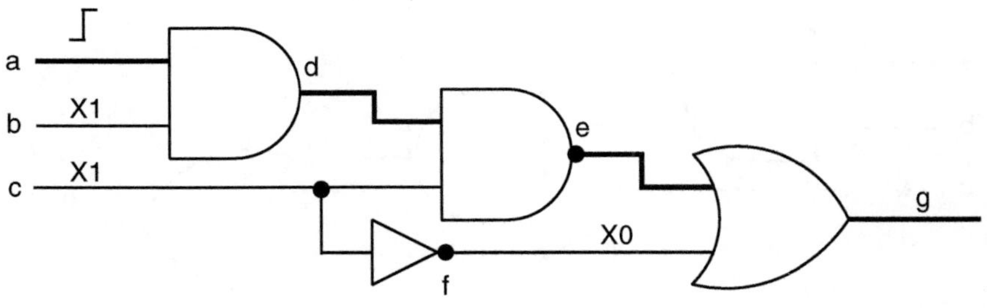

Figure 2.4: Example for non-robust sensitization

to be severe enough to affect the timing at primary outputs regardless of the timing slack along the propagation path. Further differentiations are made by whether the longest propagation path or any propagation path is considered. As in the PDFM, there always exists one fault with respect to the rising transition and one fault with respect to the falling transition. Moreover, the concepts of robust and non-robust tests are typically extended to the other delay fault models.

In this book, the *Transition Delay Fault Model* (TDFM) [104] is considered as a more relaxed fault model. Here, the delay on a single signal line is assumed to be large enough to exceed the timing slack along any propagation path to the output. Thus, gross local faults and faults distributed across a large area are covered by this fault model. Assuming m signals in the circuit, $2 \cdot m$ *Transition Delay Faults* (TDFs) can be modeled, i.e. the number of TDFs is the same as the number of stuck-at faults. But test generation for TDFs is more complex than test generation for stuck-at faults, because two time frames have to be considered.

Two test vectors v_1, v_2 are needed to test for a TDF. The vector v_1 is the test vector for timeframe t_1 and the vector v_2 the test vector for timeframe t_2. The initial vector sets the initial value of the transition at the fault site, whereas the final vector causes the transition at speed. To detect a fault, the transition must be propagated to an output by sensitizing a path from the fault site to a primary output.

Several methods have been proposed to enhance the quality of tests for TDFs. In [54], the *As-Late-As-Possible Transition Fault* (ALAPTF) model was introduced. Here, the transition is launched as late as possible to detect small delay defects. The algorithm proposed in [72] uses timing information to launch the transition and to propagate the fault effect through the longest path (timing-aware ATPG).

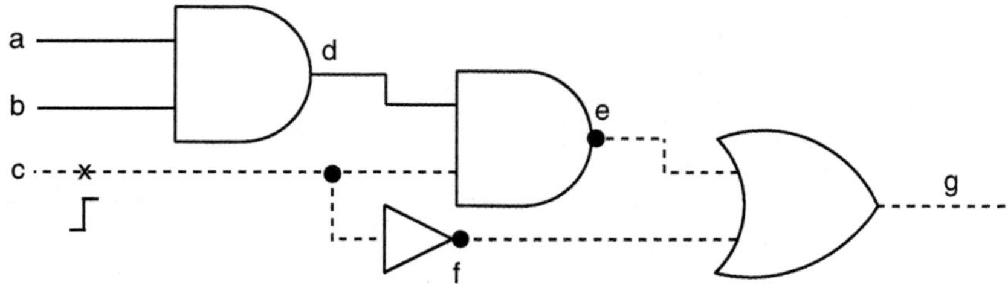

Figure 2.5: Transition delay fault example

Example 4 *Consider the circuit shown in Figure 2.5. For a test for a rising TDF at input c, two propagation paths* P_1, P_2 *are possible. They are denoted by dashed lines. The TDF could be propagated either through path* $P_1 = (c, e, g)$ *or through* $P_2 = (c, f, g)$. *Although all tests sensitizing one of the two paths are valid TDF tests, those tests sensitizing a longer path are generally preferable. This is because it is more likely that a delay defect is detected on a longer path than on a shorter path.*

2.3 Simple ATPG Framework

Figure 2.6 repeats the simple ATPG framework already shown in the introduction. As explained, *random pattern generation*, *fault simulation* and *deterministic pattern generation* are the main algorithms used. The loop containing *random pattern generation* and *fault simulation* uses efficient algorithms to find test patterns for faults that are easily testable.

In practice, this often is the majority of faults which is intuitively clear. Having a fault in a circuit that is highly optimized will more often affect the behavior of the circuit. On the other hand, some components of the circuit may only be needed to execute special functions that can only be excited under certain conditions, i.e. certain input stimuli. But not testing these functions – that may be vital to a systems safety – is not acceptable. Therefore, a deterministic engine is needed to finally decide, whether the presence of the remaining unclassified faults can influence the circuit's behavior or whether these faults are untestable.

Random pattern generation simply chooses values for all primary inputs to be applied during fault simulation. If the circuit structure is known, the probability of choosing a 0 or 1 may be adjusted. Often, however, both values are chosen with the same probability, since an exact probabilistic

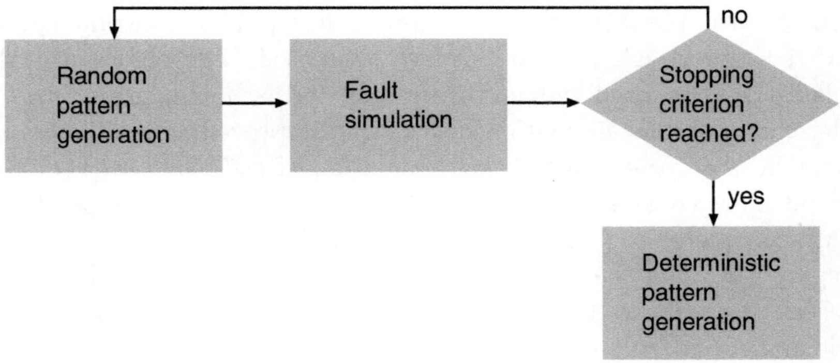

Figure 2.6: A simple ATPG framework

analysis is too costly. Only constraints coming from test compression logic are considered to adjust the probabilities. In both cases, the time to create a random pattern is linear in the number of primary inputs.

Fault simulation is also very efficient – linear in the number of gates. The simplest algorithm for fault simulation – also mentioned in the introduction – simulates a correct model of the circuit and a faulty model. If the two models yield different simulation results, the simulated pattern is a test vector. Decreasing the theoretical worst case complexity is not possible, but some techniques have been proposed that lead to an improvement in practice.

Parallel fault simulation exploits the bit-width of the data word on the underlying architecture. The arithmetic logic unit of a typical processor supports bit-wise operations on two data words. Let $a = a_1 \dots a_n$ and $b = b_1 \dots b_n$, be two bit-vectors, where n is the width of a data word. Only a single instruction is required to compute the bit-wise AND, e.g. using C-operators $c = a \;\&\; b$. Therefore, 64 test vectors can be simulated in parallel on a 64-bit machine which includes most of today's standard CPUs.

Simulating gates other than standard Boolean gates like AND, OR etc. requires additional calls. For example, tri-state buffers may be in a high impedance state. For such elements, a different encoding (see also Chapter 6) is required. In this case, more than one bit is used for the output. Still a significant speed-up can be achieved. Some additional overhead results from "packing" the test vectors into a single data word and "unpacking" them for further processing.

Besides simulating patterns in parallel, fault effects can be considered in parallel. This requires modifications to the circuit model. Additional logic is inserted to inject faults at certain gates, where a bit-mask is used to decide into which position of the bit-vector the fault is injected.

All of these parallel simulation approaches yield a constant speed-up, provided that enough test vectors are simulated. The algorithms always calculate the value of all gates in the circuit for each simulation run.

Here, event-based simulation provides an improvement. All gates in the circuit hold a current value. Upon simulation of the next test vector, only those gates are considered where the value changes. When similar test vectors are considered sequentially, often only a few value changes have to be propagated through the circuit. But still, some modifications to the circuit model are necessary to decide whether a particular fault is detected, e.g. one circuit model per fault – a huge overhead.

Deductive fault simulation decreases this effort: each gate holds the information as to which faults can be observed at its output [4]. When a value changes, this information may also change and accordingly the information about observed faults is updated.

Example 5 *Consider the circuit shown in Figure 2.7a for the input assignment $a = 0, b = 0$. The value at each signal is annotated next to the output of the gate driving the signal. Note, that there is a fanout directly after primary input b, but for simplicity, only faults at the stem of this net are considered.*

Under the assignment $a = 0, b = 0$, the stuck-at faults $(a, \text{s-a-1})$ and $(b, \text{s-a-1})$ can be observed, since both wires carry the value 0. But the effects of these faults do not propagate across AND gate c. Therefore, only fault $(c, \text{s-a-1})$ can be observed at the output of c. However, the fault $(b, \text{s-a-1})$ propagates along d that carries the value 1 in the fault free case. Thus, faults $(b, \text{s-a-1})$ and $(d, \text{s-a-0})$ can be observed at d.

The faults would switch the controlling value 1 at the input of e to 0, while the other input remains at 0 – the faults propagate across e. Consequently, both faults are observable at the primary output. Fault effects arriving at the upper input of e are canceled due to the controlling 1 at the lower input. Leading to the faults $(b, \text{s-a-1}), (d, \text{s-a-0})$ and $(e, \text{s-a-0})$ at the primary output.

Assume that the input assignment $a = 1, b = 0$ is considered next as shown in Figure 2.7b. Event-based simulation only considers primary input a. The value change does not cause any further values to flip. The faults observable at a and at c also change. The faults observable at b become observable at c as well. But now the fault $(b, \text{s-a-1})$ is observable at both inputs of e. Consequently, both effects cancel each other and the fault cannot be detected at e with the new test vector.

Even though event-based simulation and deductive fault simulation may update the value of all gates in the circuits in the worst case, often significantly fewer operations are required in practice – leading to a speed-up.

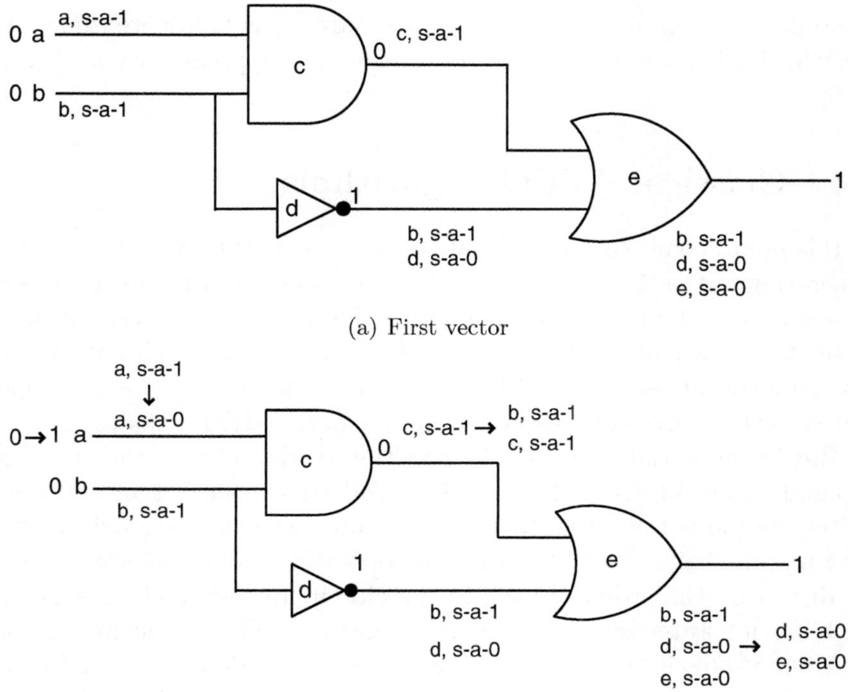

(a) First vector

(b) Next vector

Figure 2.7: Example for deductive fault simulation

On the other hand, some overhead is required. Event-based simulation requires the storing of events before propagating them through the circuit. Deductive fault simulation needs large dynamic data structures to store information about fault observation. Also, updating these data structures requires set operations that may be expensive. Finally, the sets of observable faults may be quite large, i.e. the size of the circuit in the worst case. Thus, deductive fault simulation may be quite inefficient in practice for large circuits and the large numbers of faults that must be considered.

Numerous additional techniques are needed to create a powerful simulation engine [68]. Of course, a high engineering effort is required to tune the underlying data structures for fault simulation [80]. Additionally, the extension to fault models other than simple stuck-at faults has been studied intensively, see e.g. [38, 104].

Typically more than one fault simulator is used for an efficient ATPG framework. This is due to particular needs in the ATPG flow. In the beginning where a large number of faults and random test patterns are considered, a parallel fault simulator is useful. When only a few faults are targeted potentially having similar test vectors, deductive approaches may be more useful.

2.4 Classical ATPG Algorithms

To this point, combinational circuits have been considered in the examples. Generating test patterns for circuits that contain state elements like flip-flops is computationally more difficult because the state elements cannot directly be set to a particular value. Instead the behavior of the circuit over time has to be considered during ATPG. A number of tools have been proposed that directly address this sequential problem, e.g. HITEC [83].

But in industrial practice, the resulting model often is too complex to be handled by ATPG tools. Therefore, *full scan mode* is typically applied for testing industrial designs. Here, the state elements are partitioned into several scan chains [34, 108]. In normal operation mode, the state elements are driven by the ordinary logic in the circuit. In test mode, a scan chain combines its state elements into shift registers. This allows the placing of arbitrary values into the state elements in test mode and to read out the response values after applying a test pattern. As a result, the state elements can be considered as primary inputs and outputs for testing purposes and a combinational problem results.

2.4.1 Stuck-at Faults

A symbolic formulation in terms of the Boolean difference was already considered in Section 2.2.1. Early ATPG approaches used this formulation to derive all test patterns for a specific fault. But manipulating the complex Boolean expressions resulting from this approach is typically too complex for large circuits. Later implementations based on *Binary Decision Diagrams* (BDDs) [13] allowed this approach to be applied to larger circuits. Nonetheless, these techniques need too much run time and their memory consumption is too high for practical circuits. Moreover, usually only a single test pattern is needed instead of all test patterns. Thus, more efficient techniques have been developed.

Classical algorithms for ATPG usually work directly on the circuit structure to solve the ATPG problem for a particular fault. Some of these algorithms are briefly reviewed in the following. For an in-depth discussion, the reader is referred to text books on ATPG, e.g. [14, 59].

One of the first complete algorithms dedicated to ATPG was the *D-algorithm* proposed by Roth [86]. The basic ideas of the algorithm can be summarized as follows:

- An error is observed due to differing values at a line in the circuit with or without failure. Such a divergence is denoted by values D or $\overline{D}$ to mark differences 1/0 or 0/1, respectively. (In the following, the term *D-value* means D or $\overline{D}$.)
- Instead of Boolean values, the set $\{0, 1, D, \overline{D}\}$ is used to evaluate gates and to carry out implications.
- A gate that is not on a path between the error and any output never has a D-value.
- A necessary condition for testability is the existence of a path from the error to an output, where all intermediate gates either have a D-value or are not assigned yet. Such a path is called a potential D-chain.
- A gate is on a D-chain if it is on a path from the error location to an output and all intermediate gates have a D-value.

On this basis, an ATPG algorithm focuses on justifying a D-value at the fault site and propagating this D-value to an output as shown in Figure 2.8. The algorithm starts by injecting the D-value at the fault site. Then, this value has to be propagated towards the outputs. For example, to propagate the value D at one input across a 2-input AND gate, the other input must have the non-controlling value 1. After reaching an output, the search proceeds towards the inputs in the same manner to justify the D-value at the fault site.

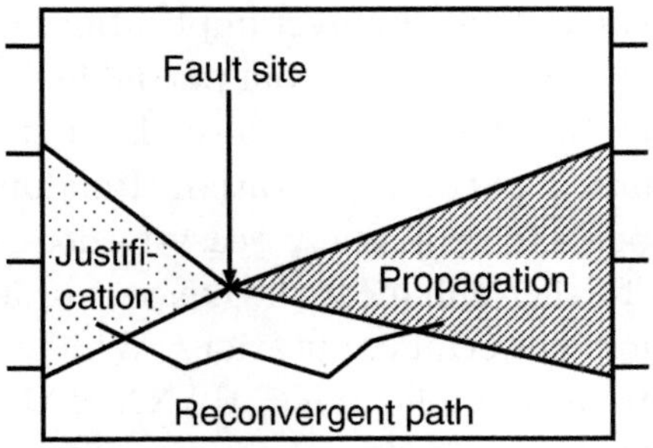

Figure 2.8: Justification and propagation

At some stages in the search, decisions are possible. For example, to produce a 0 at the output of an AND gate either one or both inputs can have the value 0. Such a decision may be wrong and may lead to a conflict later on. For example, due to a reconvergence as shown in Figure 2.8, justification may not be possible due to conditions from propagation. In this case, a backtrack search has to be applied. In summary, the D-algorithm is applied to a search space of $O(2^s)$ for a circuit with s signals including inputs, outputs and internal signals. Here the search space is defined over the number of possible decisions.

A number of improvements have been proposed for this basic procedure. PODEM [48] branches only on the values for primary inputs. This reduces the search space for test pattern generation to $O(2^n)$ for a circuit with n primary inputs. But as a disadvantage, time is wasted if all internal values are implied from a given input assignment that finally does not detect the fault.

FAN [44] improves upon this problem by branching on stems of fanout points as well. This incorporates internal structure of the circuit in the computation of a test pattern. The branching order and value assignments are determined by heuristics that rely on controllability and observability measures (e.g. SCOAP [51]) to predict a "good" variable assignment for justification or propagation, respectively. Moreover, the algorithm keeps track of a justification frontier moving towards the inputs and a propagation frontier moving towards the outputs. Therefore, FAN can make the "most important decision" first – based on a heuristic – while the D-algorithm applied a static order doing only propagation at first and justification afterwards.

SOCRATES [88] includes the use of global static implications by considering the circuit structure. Based on particular structures in the circuit, indirect implications are possible, i.e. implications that are not directly obvious due to assignments at a single gate, but rather result from functional arguments across several gates. These indirect implications are applied during the search process to imply values earlier from partial assignments and, by this, prevent bad decisions. SOCRATES was also the first approach combining random and deterministic pattern generation. Random pattern generation is applied as a preprocess to drop many "easy-to-detect" faults.

HANNIBAL [65] further enhances the concept of learning from the circuit structure with more powerful implications. While SOCRATES only uses a predefined set of indirect implications, HANNIBAL learns from the circuit structure. For this task, recursive learning [66] is applied. In principle, recursive learning is complete by itself, but too time consuming to be used as a stand alone procedure. Therefore, learning is done in a preprocessing step, during which the effects of value assignments are calculated and the resulting implications are learned. These implications are stored for the following run of the search procedure. In HANNIBAL, the FAN algorithm was used to realize the search step.

The algorithms introduced so far work on a structural description, i.e. a netlist, of the circuit. IGRAINE [97] introduces a new *Implication Graph* (IG) model. The IG represents the logic function as well as the topology of the circuit in a single model of the CNF. By this, the time consuming task of justification and propagation can be reduced to significantly simpler graph algorithms. Furthermore, the algorithms are more flexible and can be applied to IGs derived from different logic systems.

SPIRIT [47] also works on an IG and introduces a new, more efficient data structure for the complete IG. Furthermore, a large number of standard ATPG techniques, e.g. X-path check [48] and unique sensitization [44] have been ported to the IG model. Static learning [88] as well as recursive learning [66] are integrated.

Due to the use of an IG, both IGRAINE and SPIRIT are SAT-based algorithms and benefit from the unified data structure of the graph model. However, both approaches work differently to the SAT-based approach presented in this book. Instead of using a SAT solver as a black box (as proposed in this book), specific routines for justification and propagation are applied in IGRAINE as well as in SPIRIT.

2.4.2 Delay Faults

Four different classes of ATPG algorithms for PDFs can be identified: structure-based, algebraic, non-enumerative and SAT-based algorithms. Structure-based ATPG algorithms work directly on the circuit structure performing path sensitization and value justification for each targeted path. Several logic systems were proposed that can handle different sensitization criteria and provide efficient implication procedures. For example, Figure 2.9 shows the Hasse diagram of the ten-valued logic proposed in [43]. The lowest level shows the basic values of the logic and the upper levels present the composite values.

The approach in [71] uses a five-valued logic to generate test patterns and introduces the general robust sensitization criterion. DYNAMITE [41] proposes a ten-valued and a three-valued logic system for robust and non-robust test generation, respectively. It provides a stepwise path sensitization procedure which is capable of proving large numbers of paths as untestable by a single test generation attempt. Consequently, DYNAMITE

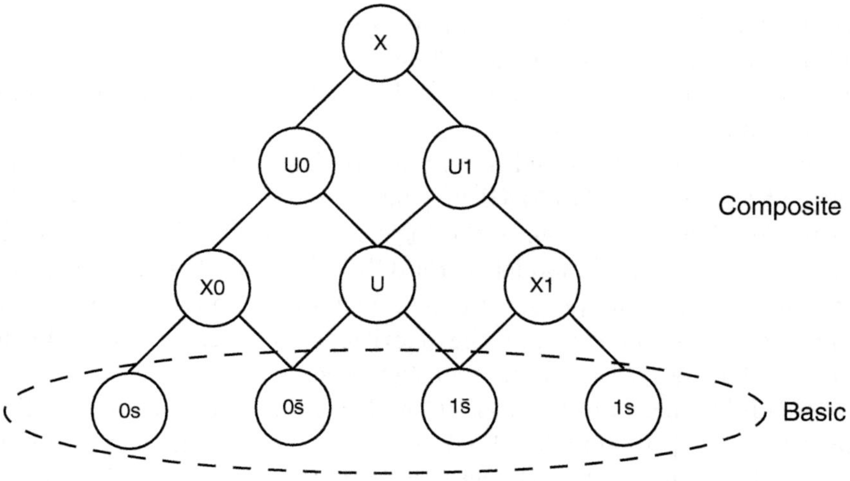

Figure 2.9: Hasse diagram of ten-valued logic [43]

is very effective for circuits with a large number of untestable PDFs. The approach in [43] enhances this scheme by using five different logic systems, e.g. a ten-valued and a 51-valued logic system, suitable for various test classes such as non-robust, robust and restricted robust.

Algebraic algorithms do not work on the circuit structure, but on Boolean expressions, e.g. represented as (RO)BDDs. The approach in [5] converts the circuit and the constraints that have to be satisfied for a delay test to BDDs. For each fault, a pair of constraints is considered. Each constraint corresponds to one of the two time frames. Robust as well as non-robust tests are then obtained by evaluating the BDDs.

The tool BiTeS [24] constructs BDDs for the strong robust PDFM, i.e. for generating hazard-free tests. Instead of generating one single test pattern for a PDF, the complete set of tests is generated directly using BDDs. Thus, procedures for test set compaction can easily be applied. However, BDD-based algorithms suffer from the large size of the constructed BDDs. The memory consumption of the BDDs are not feasible for industrial circuits.

Non-enumerative ATPG algorithms do not target any specific path, but generate tests for all PDFs in the circuit. Hence, the problem of the exponential number of paths in a circuit is avoided. The first non-enumerative ATPG algorithm was NEST [85]. NEST considers all single lines in the circuit rather than the exponential number of paths. This greedy approach is based on propagating transitions robustly through parts of the circuit, i.e. sub-circuits. For selected sub-circuits, i.e. sub-circuits with a large number of structurally compatible paths, test objectives are determined and tests are generated. Two paths are structurally compatible if they have the same endpoint and the same parity. Fault simulation is done for each test, since each test potentially detects many faults. The approach is most effective in highly testable circuits. Due to the greedy nature, the approach does not perform well on poorly testable circuits.

The tool ATPD [101] improves the techniques from NEST by proposing a new sensitization phase. While NEST is only based on sub-circuits that have many structurally compatible paths, ATPD takes the robust sensitization criterion into account for the selection of sub-circuits. As a result, tests generated within these sub-circuits detect more PDFs.

RESIST [42] exploits the fact that PDFs are dependent, because many paths share sub-paths. Therefore, RESIST does not enumerate all possible paths, but sensitizes sub-paths between two fanouts, between input and fanout, or between fanout and output, respectively. The approach sensitizes each sub-path only once and, consequently, decreases the number of sensitization steps.

SAT-based algorithms work differently from those presented in this section. The problem of generating a test for a PDF is transformed to a Boolean SAT problem. The SAT problem is solved by a SAT solver. The SAT solution is then transformed into a solution of the original problem, i.e. a PDF test pattern. A detailed description of the basic SAT concepts is given in Chapter 3. The first SAT-based approach for PDF test generation was proposed in [17] where a seven-valued logic is used to generate robust tests for PDFs in combinational circuits. For the transformation into a Boolean SAT problem, a Boolean encoding is applied.

In [61], *Incremental SAT* [55, 92] is used to speed up non-robust PDF test generation. Similar to [41], paths are incrementally sensitized causing a large number of untestable path delay faults to be pruned in a single step. This procedure is enhanced by the approach presented in [16]. Here, unsatisfiable cores are generated for untestable paths to identify sub-paths that cannot be sensitized. Other paths containing these sub-paths can directly be marked as untestable, too.

The tool KF-ATPG is presented in [109]. Unlike the above mentioned SAT-based approaches, KF-ATPG uses the circuit-based SAT solver presented in [74]. Therefore, it is able to exploit structural knowledge of the problem to speed up test generation. Furthermore, KF-ATPG works on a path-status graph to keep track of testable and untestable paths. However, this approach is limited to non-robust test generation. The underlying circuit-based SAT solver works only on Boolean logic.

All of these approaches based on SAT show the potential of SAT-based ATPG for delay faults. However, the SAT-based ATPG algorithms for PDFs mentioned thus far either are not able to generate high quality tests, i.e. robust tests, or do not model the sequential behavior of the circuit adequately. These problems are addressed in Chapter 10 where test pattern generation is considered for delay faults in large industrial circuits.

2.5 Benchmarking

The research work presented in this book is experimentally evaluated. This section provides general information on how the experiments are conducted. The experiments were carried out on publicly available benchmark circuits, i.e. ISCAS'85 [12], ISCAS'89 [11] and ITC'99 [20] benchmark suites, as well as on industrial circuits.

Tables 2.2 and 2.3 show statistical information about the industrial circuits provided by NXP Semiconductors, Hamburg, Germany. Statistical information about the publicly available benchmark circuits is easy to obtain and therefore not reported here. All industrial circuits have been found to be difficult cases for test pattern generation. Table 2.2 gives input/output information. In the first column, the name of the industrial circuit is given.

Table 2.2: Statistical information about industrial circuits – input/output

Circuit	PI	PO	FF	IO	Fixed
p44k	739	56	2,175	0	562
p49k	303	71	334	0	1
p57k	8	19	2,291	0	8
p77k	171	507	2,977	0	2
p80k	152	75	3,878	0	2
p88k	331	183	4,309	72	200
p99k	167	82	5,747	0	102
p141k	789	1	10,507	0	790
p177k	768	1	10,507	0	769
p456k	1,651	0	14,900	72	1,655
p462k	1,815	1,193	29,205	0	1,458
p565k	964	169	32,409	32	861
p1330k	584	90	104,630	33	519
p2787k	45,741	1,729	58,835	274	45,758
p3327k	3,819	0	148,184	274	4,438
p3852k	5,749	0	173,738	303	5,765

Table 2.3: Statistical information about industrial circuits – size

Circuit	Gates	Fanouts	Tri	Targets
p44k	41,625	6,763	0	64,105
p49k	48,592	14,323	0	142,461
p57k	53,463	9,593	13	96,166
p77k	74,243	18,462	0	163,310
p80k	76,837	26,337	0	197,834
p88k	83,610	14,560	412	147,048
p99k	90,712	19,057	0	162,019
p141k	116,219	25,280	560	267,948
p177k	140,516	25,394	560	268,176
p456k	373,525	94,669	6,261	740,660
p462k	417,974	75,163	597	673,465
p565k	530,942	138,885	17,262	1,025,273
p1330k	950,783	178,289	189	1,510,574
p2787k	2,074,410	303,859	3,675	2,394,352
p3327k	2,540,166	611,797	44,654	4,557,826
p3852k	2,958,676	746,917	22,044	5,507,631

Note, that the name of the circuit roughly denotes the size of the circuit, e.g. p3852k contains over 3.8 million elements.

Column *PI* presents the number of primary inputs of the circuit. The number of primary outputs is given in column *PO*. In column *FF*, the number of state elements, i.e. Flip-Flops, is shown, while column *IO* denotes the number of Input/Output elements. Column *Fixed* gives the number of (pseudo) primary inputs which are restricted to a fixed value. These inputs are not fully controllable. More information about the reasons for and consequences of these fixed inputs can be found in Chapter 6.

Table 2.3 shows further information about the circuits' internals. Column *Gates* gives the number of gates in the circuit, while the column *Fanouts* presents the number of fanout elements. Further, in column *Tri*, the number of tri-state elements, e.g. bus, busdrivers, is given. Column *Targets* shows the number of stuck-at fault targets, i.e. the number of stuck-at faults that have to be tested. If not noted otherwise, fault dropping is activated during test generation. When a test pattern is generated, the fault simulation detects all other faults that can be detected by this pattern. All of these detected faults are dropped from the fault list. No expensive test generation is required for the dropped faults.

The SAT-based ATPG tool PASSAT, which incorporates the techniques presented in this book, has been integrated as a prototype into the ATPG framework of NXP Semiconductors. This framework has been developed at Philips Semiconductors and later at NXP Semiconductors for more than 25 years. The experiments were performed over a period of more than 4 years. Therefore, the experimental setup varies between the reported experiments.

The SAT-based ATPG tool PASSAT has been improved in steps. Furthermore, the ATPG framework itself has been improved during this period. This also influences the experiments and, as a consequence, results for a particular circuit may differ slightly throughout the book. However, experimental results presented in a single section are consistent, i.e. carried out on the basis of the same framework and on the same machine.

In the first experiments, the SAT solver zChaff [8, 82] was used as basic SAT engine, whereas in later experiments, MiniSat version 1.14 [28, 29] was applied to test generation. The underlying SAT engine was changed due to the high performance gain which was achieved by using MiniSat v1.14. Tests with the newer version, MiniSat 2, were also executed. However, this consistently resulted in an increased run time. Therefore, MiniSat v1.14 was kept as the SAT engine.

We also experimentally evaluated the publicly available circuit-based SAT solver CSAT [74, 75] (downloadable from [73]; for more information about circuit-oriented SAT solvers, see Section 3.4). However, the CNF-based SAT solver MiniSat turned out to be faster in solving ATPG problems.

There are three different states for each target fault: testable, untestable and aborted. A target fault is *testable* if a test that detects the fault has been generated. A fault is *untestable* if it is proven that no such test exists. If neither has been concluded within a given interval, the target fault is considered to be *aborted*. Two different types of intervals were used: time interval and restart interval. The time interval aborts test generation for a target fault after a given number of CPU seconds. In the standard experimental setup, test generation is aborted after 20 CPU seconds for a specific fault.

The restart interval aborts test generation after a given number of restarts of the SAT solver (see Section 3.2.5). A restart is triggered after a certain number of conflicts (100 at the beginning in MiniSat). After each restart, the number of permissible conflicts is increased (by 50% in MiniSat). Defining the interval using restarts has the advantage of being machine-independent. The standard value is 12 restarts. In MiniSat, this corresponds to 25,650 conflicts – note the use of integer arithmetic.

Information about the concrete setup used for each experiment, such as the machine used or the search parameters, is provided in the corresponding sections.

Chapter 3

Boolean Satisfiability

This chapter provides the background on *Boolean Satisfiability* (SAT) necessary for the description, in the next chapter, of an ATPG engine based on SAT. The main concepts of SAT solvers are reviewed in Section 3.1. An overview of advanced SAT techniques is given in Section 3.2. In Section 3.3, the standard transformation of a given circuit into constraints for a SAT solver is introduced. Finally, Section 3.4 provides a short overview of circuit-oriented SAT techniques.

3.1 SAT Solver

SAT solvers usually work on a database that represents the Boolean formula in *Conjunctive Normal Form* (CNF) or product of sums. A CNF is a conjunction (product) of clauses, where each clause is a disjunction (sum) of literals. A literal is a variable or its complement.

Example 6 *The following Boolean formula is given in CNF:*

$$\Phi = \underbrace{(a + \overline{c} + d)}_{\omega_1} \cdot \underbrace{(a + \overline{c} + \overline{d})}_{\omega_2} \cdot \underbrace{(a + c + d)}_{\omega_3} \cdot \underbrace{(a + c + \overline{d})}_{\omega_4} \cdot \underbrace{(\overline{a} + \overline{b} + c)}_{\omega_5}$$

Upper case Greek letters $\Omega, \Phi, \Xi, \ldots$ denote CNF formulas; the lower case Greek letter $\kappa, \omega_1, \omega_2, \ldots$ denote clauses and $\lambda_1, \lambda_2, \ldots$ denote literals. Variables are denoted by lower case Latin letters $a, b, \ldots$.

Remark 1 *Unless stated otherwise, the above notation is used for CNF formulas. Alternatively, a CNF can be considered as a set of clauses and a*

R. Drechsler et al., *Test Pattern Generation using Boolean Proof Engines*,

clause as a set of literals. In some cases, the set-based notation is more convenient. In the following, the set-based notation is only used when explicitly stated.

The objective during SAT solving is to find a satisfying assignment for the given Boolean formula or to prove that no such assignment exists. A CNF is satisfied if all clauses are satisfied. A clause is satisfied if at least one literal in the clause is satisfied. The positive literal a is satisfied if the value 1 is assigned to variable a. The negative literal $\overline{a}$ is satisfied if the value 0 is assigned to variable a. Besides being satisfied, a clause can assume two further states. A clause is unsatisfied under the current (partial) assignment if all of its literals are assigned negatively. If a clause is neither *satisfied* nor *unsatisfied*, i.e. none of its literals is assigned positively and there is at least one free literal, the clause is called *not satisfied*.

Modern SAT solvers are based on the DPLL procedure that was first introduced by Davis and Putnam [22] and was improved by Davis, Logeman and Loveland [21]. Often the DPLL procedure is also referred to as DLL. In principle, this algorithm explores the search space of all assignments by a backtracking search as described in Figure 3.1.

To begin, a decision is made by choosing a variable and a value for this variable according to a *variable selection strategy* (Step 1). Then, implications due to this assignment are carried out (Step 2). When all clauses are satisfied, the problem is solved (Step 3). Otherwise, the current assignment may only be partial and therefore, no conclusion is possible yet. In this case, further assignments are necessary (Step 4).

If at least one clause is not satisfied under the current (partial) assignment, *conflict analysis* is carried out (Step 5) as will be explained in more detail in Section 3.2. Then, a new branch in the search tree is explored by inverting the value of the variable. This is also known as *flipping* (Step 6). When there is no decision to undo, the search space has been completely explored and therefore, the instance is unsatisfiable (Step 7).

3.2 Advances in SAT

SAT solvers have become a powerful engine to solve real world problems only after some substantial improvements to the basic DPLL procedure in the recent past. These improvements include efficient *Boolean Constraint Propagation* (BCP), *conflict analysis* together with *non-chronological backtracking* and sophisticated *variable selection strategies*.

1. *Decision*:
   Choose an unassigned variable and assign a new value to the variable.
2. *Boolean Constraint Propagation*:
   Carry out implications resulting from the assignment chosen in Step 1.
3. *Solution*:
   If all clauses are satisfied, output the current variable assignment and return “satisfiable”.
4. If there is no unsatisfied clause due to the current assignment proceed with Step 1.
5. *Conflict analysis*:
   If the current assignment leads to at least one unsatisfied clause, carry out conflict analysis and add a conflict clause.
6. *(Non-chronological) Backtracking*:
   Undo the most recent decision, where switching the variable could lead to a solution, undo all implications due to this assignment and switch the variable value. Goto Step 2.
7. *Unsatisfiable*:
   Return “unsatisfiable”.

Figure 3.1: DPLL procedure

3.2.1 Boolean Constraint Propagation

BCP carries out implications due to previous decisions. In order to satisfy a CNF, all clauses must be satisfied. Assume that under the current partial assignment all but one of the literals in a clause evaluate to 0 and the variable of the last literal is unassigned. Then, the value of this last variable can be implied in order to satisfy the clause.

Example 7 *Again, consider the CNF from Example 6. Assume the partial assignment $a = 1$ and $b = 1$. Then, clause ω_5 implies the assignment $c = 1$.*

BCP has to be carried out after each variable assignment and the efficiency of this procedure is thus crucial to the overall performance. In [82], an efficient data structure for BCP was presented for the SAT solver Chaff (the source code of the implementation zChaff can be downloaded from [8]).

The basic idea is to use the *two literal watching scheme* to efficiently detect, when an implication may be possible. Two literals of each clause are watched. An implication can occur for the clause only if one of these literals evaluates to 0 upon a previous decision and the other literal is unassigned. If, due to a second unassigned literal no implication occurs, this second literal is watched.

For each literal, a watching list is stored to efficiently access those clauses where the particular literal is watched. Therefore, instead of always looping over all clauses or all clauses containing the assigned variable in the database, only those that may cause an implication are considered.

3.2.2 Conflict Analysis

Conflict analysis was first proposed in [79] for the SAT solver GRASP. In the traditional DPLL procedure, only the most recent decision was undone, when a conflict, i.e. a clause that is unsatisfied under the current assignment, was detected. In contrast, modern SAT solvers analyze such a conflict. During BCP, a conflict occurs if opposite values are implied for a single variable due to different clauses. Then, the reasons (i.e. the assignments) that are responsible for this conflict are detected and a conflict clause is generated from these reasons.

Conflict clauses are logically redundant and the size of the SAT instance grows dynamically. As an advantage, these clauses represent illegal partial assignments and, by this, prevent the solver from re-entering the same non-solution search space again. In contrast, the traditional DPLL algorithm without conflict analysis does not save the information and would therefore enter this particular search space again under a different partial assignment after backtracking.

For the creation of conflict clauses, an *implication graph* is maintained during the search that keeps track of the reasons for each assignment. Each node in the implication graph represents an assignment. Decisions are represented by nodes without predecessors. Each implied assignment has the reason that caused the assignment at its predecessors. The edges are labeled by the clauses that cause the assignment. An example of an implication graph is shown in Figure 3.2.

Finally, *non-chronological backtracking* is performed, i.e. the SAT solver backtracks to the decision before the last decision that participated in the conflict. Flipping the value of the last decision that led to the conflict is done by BCP due to the inserted conflict clause (*failure-driven assertion*). Therefore, this value assignment becomes an implication instead of a decision. The following example demonstrates the procedure.

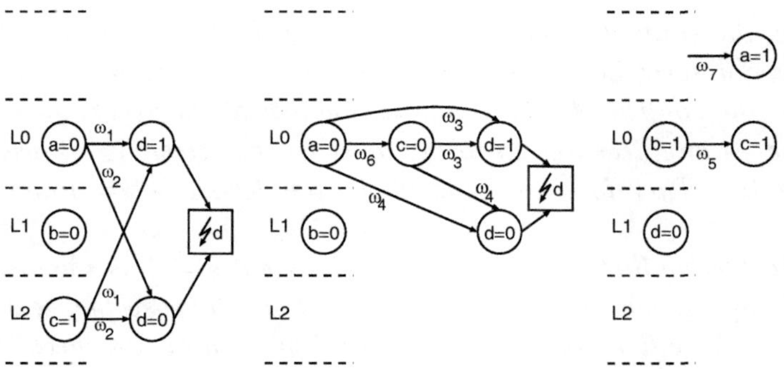

(a) Snap shot 1 (b) Snap shot 2 (c) Snap shot 3

Figure 3.2: Decision stack

Example 8 *Consider the CNF* Φ

$$\Phi = \underbrace{(a+\overline{c}+d)}_{\omega_1} \cdot \underbrace{(a+\overline{c}+\overline{d})}_{\omega_2} \cdot \underbrace{(a+c+d)}_{\omega_3} \cdot \underbrace{(a+c+\overline{d})}_{\omega_4} \cdot \underbrace{(\overline{a}+\overline{b}+c)}_{\omega_5}$$

(from Example 6). Each time the SAT solver makes a decision, this decision is pushed onto the decision stack. *Now, assume that the first decision at* decision level *L0 is the assignment* $a = 0$. *No implications follow from this decision. Then, at L1, the solver chooses* $b = 0$. *Again, no implications follow. At L3, the solver chooses* $c = 1$. *Now, clause* ω_1 *implies the assignment* $d = 1$, *but* ω_2 *implies* $d = 0$. *Therefore, a conflict with respect to variable d occurs.*

This situation is shown in Figure 3.2a. The decision stack is shown on the left hand side. The solver tracks reasons for assignments using the implication graph shown in the right part of Figure 3.2a. In this example, the assignments $a = 0$ *and* $c = 1$ *caused the assignment* $d = 1$ *due to clause* ω_1. *Additionally, this caused the assignment* $d = 0$ *due to* ω_2 *and a conflict results. By traversing the graph backwards, the reason for the conflict, i.e.* $a = 0$ *and* $c = 1$, *can be determined.*

Now, it is known that this assignment must be avoided in order to satisfy the CNF. This information is stored by adding the conflict clause $\omega_6 = (a+\overline{c})$ *to the CNF. Thus, the same non-solution space is never re-entered during further search – this is also called* conflict-based learning. *The decision* $c = 1$ *is undone. Due to* $a = 0$ *and the conflict clause* ω_6 *the assignment* $c = 0$ *is implied which is called a* failure driven assertion.

The implication $c = 0$ *triggers a next conflict with respect to d as shown in Figure 3.2b. The single reason for this conflict is the decision* $a = 0$.

Therefore, the conflict clause $\omega_7 = (a)$ *is added. Now, the solver backtracks above decision level L0. This happens because the decision* $b = 0$ *was not a reason for the conflict. Instead, non-chronological backtracking occurs – the solver undoes any decision up to the most recent decision that was involved in the conflict. Therefore, in the example, the decisions* $b = 0$ *and* $a = 0$ *are undone.*

Due to the conflict clause ω_7*, the assignment* $a = 1$ *is implied independently of any decision as shown in Figure 3.2c. Suppose the next choice at L0 is* $b = 1$*. For efficiency reasons, the SAT solver does not check whether all clauses are satisfied under this partial assignment, but only detects conflicts. Therefore, a satisfying assignment is found by deciding* $d = 0$ *at L1.*

In summary, this example showed on an informal basis how a modern SAT solver carries out conflict analysis and uses conflict clauses to "remember" non-solution spaces. Generally, the shorter the conflict clause that can be derived, the larger is the subspace that is pruned. Essentially, any cut through the implication graph that separates the conflict node from the decisions responsible for the conflict is a valid conflict clause. Heuristics are applied to derive good conflict clauses.

Here, the concept of a *Unique Implication Point* is important: only a single literal assigned on the most recent decision level is contained in the cut. This ensures, that the most recent decision is replaced by an implication together with the conflict clause. A formal and more detailed presentation of the technique can be found in [79]. The algorithms to derive conflict clauses have been further improved, e.g. in [29, 110]. Moreover, in [27], *self-subsumption* was proposed. Due to self-subsumption, conflict clauses may be reduced by resolution with clauses involved in this conflict.

A result of this learning and improved conflict clause generation is a significant speed-up of the solving process – in particular also for unsatisfiable formulas.

While learning is a conceptual improvement, adding conflict clauses also causes overhead: higher memory requirements and longer run times for BCP. Therefore, learned clauses are regularly deleted to keep this overhead acceptable. Here, the activity of clauses is a recent very efficient approach. Counters are associated to clauses and incremented when a clause is considered during conflict analysis [29, 49]. These counters are regularly divided by a given value to emphasize more recent influences. Those conflict clauses with a low activity are periodically deleted from the clause database.

3.2.3 Variable Selection Strategies

Another important improvement of SAT solvers results from sophisticated variable selection strategies. A detailed overview about the effect of decision strategies on performance can be found in [76]. Basically, the SAT solver dynamically collects statistics about the occurrence of literals in clauses. A dynamic procedure is used to keep track of conflict clauses added during the search. An important observation is that locality is achieved by exploiting recently learned information. This helps to speed up the search.

An example is the *Variable State Independent Decaying Sum* (VSIDS) strategy employed in [82]. Basically, this strategy attempts to satisfy recently learned conflict clauses. A counter exists for each literal to count the number of occurrences in clauses. Each time a conflict clause is added, the appropriate counters are incremented. The value of these counters is regularly divided by two which helps to emphasize the influence of more recently learned clauses. A large number of other heuristics has also been investigated, e.g. in [49, 60, 76].

3.2.4 Correctness and Unsatisfiable Cores

Typically, a SAT solver takes a CNF formula as input and determines whether the CNF formula is satisfiable or unsatisfiable. Verifying satisfiability of a CNF formula is trivial once a potential satisfying assignment is known. Given the solution produced by the SAT solver, an independent tool checks whether all clauses are satisfied under this assignment. The run time is linear in the number of clauses. Verifying unsatisfiability is not as obvious as verifying satisfiability. In [50, 111], methods are presented that validate the results of the SAT solvers zChaff [82] and BerkMin [49], respectively.

The main idea to proof unsatisfiability is to generate an empty clause from a sequence of resolutions among the original clauses [111] or among the conflict clauses [50]. For this, a trace is produced during the search. For each learned conflict clause, those clauses which have been responsible for the conflict (resolvent clauses) are recorded in a *Directed Acyclic Graph* (DAG). The leaf nodes of the DAG represent the original clauses. If unsatisfiability is determined, the final conflicting assignment is taken as the starting point to traverse the recorded DAG for proving unsatisfiability. Resolution is then carried out using the recorded trace, i.e. the DAG, to generate the empty clause. If no empty clause can be generated by resolution, the result of the SAT solver is not correct.

Given an unsatisfiable CNF formula Φ, a subset of clauses $\Omega \subseteq \Phi$ is called an *unsatisfiable core* if it is unsatisfiable by itself. This unsatisfiable core is often extracted to determine the reason for the unsatisfiability of a problem. In the worst case, the unsatisfiable core Ω is equal to the original CNF formula Φ.

A common method to extract an unsatisfiable core is to use the method described above to verify unsatisfiability. The set of all original clauses used to generate the empty clause is unsatisfiable by itself and therefore forms an unsatisfiable core. Minimal unsatisfiable cores are useful in many applications to "locate the reason" for unsatisfiability. But solving this optimization problem is computationally much more expensive than using a potentially larger unsatisfiable core derived from the trace. The approach presented in [70] deals with the extraction of minimal unsatisfiable cores.

3.2.5 Optimization Techniques

Today, improvement of SAT solvers is an active field of research. Besides the major improvements presented in the last sections, new SAT techniques have been developed to speed up the search process. In the following, some of the optimizations are briefly reviewed.

Restarts

Random restarts have been proposed to avoid spending too much run time in hard subspaces without solutions [82]. After a given interval, the SAT solver is restarted, resetting some of the statistical information. As a result, a different part of the search space is entered. If random restarts happen too often, the solver becomes incomplete, i.e. unsatisfiability cannot be proven. To avoid this, the intervals between random restarts are continuously increased. As a result, the SAT solver can finally exhaust the search space to proof unsatisfiability – provided that sufficient resources are available.

Recently, a new adaptive restart strategy was proposed in [6] that measures the agility of the SAT solver based on the number of flipped assignments. This avoids counterproductive restarts.

Preprocessing

Another ingredient to modern SAT solvers is a powerful preprocessing step as proposed in [25, 27, 60]. The original CNF is usually a direct mapping of the problem onto a CNF representation. No optimizations are carried out, e.g. unit clauses are frequently contained in this original CNF, but these can be eliminated without changing the solution space. Typically, all implications due to unit clauses in the initial CNF formula are carried out – this removes some variables and shortens clauses. For hard problem instances, also more elaborate and time consuming techniques pay off.

For example, subsumption is exploited e.g. in [27]: if the CNF contains clauses ω_1 and ω_2, where $\omega_1 = \omega_2 \cup a$ (and a is a set of literals), then ω_2 can be removed – whenever ω_1 is satisfied, ω_2 is also satisfied. Additionally, some simple resolution steps are often carried out to partially simplify the formula. More expensive techniques try to identify and remove symmetries in the problem instance [3]. However, these approaches are typically either too slow to be incorporated into the SAT solver or can be applied only for certain problems. In summary, when preprocessing the CNF formula, fast optimizations are applied to make the representation more compact and to improve the performance of BCP.

Structural Information

Often, the original problem formulation provides more insight into the structure of the problem. This can be exploited while generating the CNF formula. For example, additional clauses are inserted to increase the reasoning power of BCP. Alternatively, the structure of the problem may be exploited to create a small SAT instance. A very similar technique is tuning the heuristics of the SAT solver. For example, structural knowledge has been used to bypass the decision heuristic [92]. Similarly, learned clauses can be generalized. If the structure that caused a conflict can be identified, the learned clause can be replicated for other identical structures within the same SAT problem [92] or can be reused for other SAT instances that contain the same structure [77].

Parameter Settings

Overall, SAT solvers have a wide range of possible experimental settings. Moreover, different parameters such as for example restart interval and decision variable ordering interact in a complex way. State-of-the-art SAT solvers are often manually tuned on a large set of benchmarks. However, in [57], it is shown that automated parameter optimization based on genetic algorithms can increase the effectiveness of a SAT solver significantly.

Due to all of the advances explained above, SAT solvers have become the state of the art for solving a large range of problems in CAD, e.g. formal verification [7,64], debugging and diagnosis [2,35,93].

3.3 Circuit-to-CNF Conversion

A SAT solver can be applied as a powerful black-box engine to solve a problem. In this case, transforming the problem instance into a SAT instance is a critical step. In particular SAT-based ATPG require the transformation of the circuit into a CNF. The basic procedure has been presented in [67, 102].

The transformation of a single AND gate into a set of clauses is shown in Table 3.1. The goal is to create a CNF that models an AND gate, i.e. a CNF that is only satisfied for assignments that may occur for an AND gate. Let a and b be the two inputs and c the output of an AND gate, then c must always be equal to $a \cdot b$. The truth table for this CNF formula is shown in Table 3.1a. From the truth table, a CNF formula is generated by extracting one clause for each assignment where the formula evaluates to 0 and applying De Morgan's theorem. These clauses are shown in Table 3.1b. This CNF representation is not minimal and can therefore be reduced by two-level logic minimization, e.g. using SIS [89]. The clauses in Table 3.1c are the final result.

The generation of the CNF for a complete circuit is straightforward. For each gate, clauses are generated according to the type of the gate. The output variables and input variables of a gate and its successors are identical and therefore establish the overall circuit structure within the CNF. Formally, the circuit is transformed into a CNF Φ by building the conjunction of the CNFs Ω_g for all gates g:

$$\Phi = \bigwedge_{g \in C} \Omega_g.$$

Table 3.1: Transformation of an AND gate into CNF

(a) Truth table				(b) Clauses	(c) Minimized
a	b	c	$c \leftrightarrow a \cdot b$		
0	0	0	1		
0	0	1	0	$(a + b + \overline{c})$	
0	1	0	1		$(a + \overline{c})$
0	1	1	0	$\cdot\ (a + \overline{b} + \overline{c})$	$\cdot\ (b + \overline{c})$
1	0	0	1		$\cdot\ (\overline{a} + \overline{b} + c)$
1	0	1	0	$\cdot\ (\overline{a} + b + \overline{c})$	
1	1	0	0	$\cdot\ (\overline{a} + \overline{b} + c)$	
1	1	1	1		

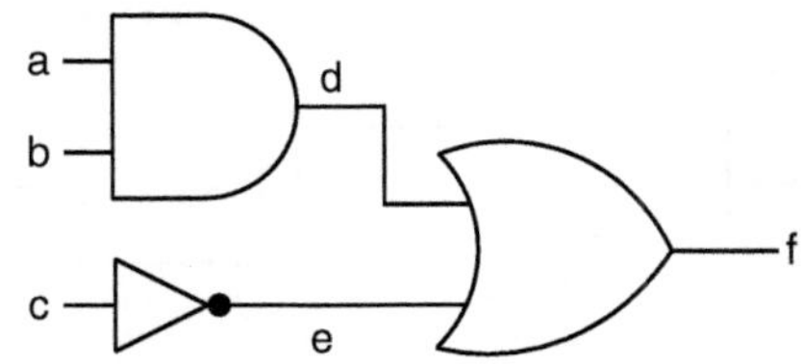

Figure 3.3: Example for transformation (replicated from Figure 2.2a)

Example 9 *Consider the circuit shown in Figure 3.3. The CNF formulas for the single gates are:*

$$\Phi_d = \underbrace{(a + \overline{d}) \cdot (b + \overline{d}) \cdot (\overline{a} + \overline{b} + d)}_{d \leftrightarrow a \cdot b}$$

$$\Phi_e = \underbrace{(c + e) \cdot (\overline{c} + \overline{e})}_{e \leftrightarrow \overline{c}}$$

$$\Phi_f = \underbrace{(\overline{d} + f) \cdot (\overline{e} + f) \cdot (d + e + \overline{f})}_{f \leftrightarrow d+e}$$

Consequently, the CNF formula for the entire circuit is given by:

$$\Phi = \Phi_d \wedge \Phi_e \wedge \Phi_f$$

An advantage of this transformation is the linear size complexity. Given a circuit where n is the sum of the numbers of inputs, outputs and gates, the number of variables in the SAT instance is also n and the number of clauses is $O(n)$.

There is some degree of freedom how to translate a given circuit into a CNF formula. Consider the following example to understand the trade-offs in this translation.

Example 10 *Consider the circuit in Figure 3.4. This is a multiplexer realized by basic gates. Obviously, if both data inputs d_0 and d_1 have the same value, the output o must assume this value, too.*

By transforming each single gate into CNF, the following CNF formula Φ, with the auxiliary variables t_0, t_1, results:

$$\begin{aligned}\Phi \quad = \quad & \underbrace{(d_0 + \overline{t_0}) \cdot (\overline{s} + \overline{t_0}) \cdot (\overline{d_0} + s + t_0)}_{t_0 \leftrightarrow d_0 \cdot \overline{s}} \\ \cdot \quad & \underbrace{(d_1 + \overline{t_1}) \cdot (s + \overline{t_1}) \cdot (\overline{d_1} + \overline{s} + t_1)}_{t_1 \leftrightarrow d_1 \cdot s} \\ \cdot \quad & \underbrace{(\overline{t_0} + o) \cdot (\overline{t_1} + o) \cdot (t_0 + t_1 + \overline{o})}_{o \leftrightarrow t_0 + t_1}\end{aligned}$$

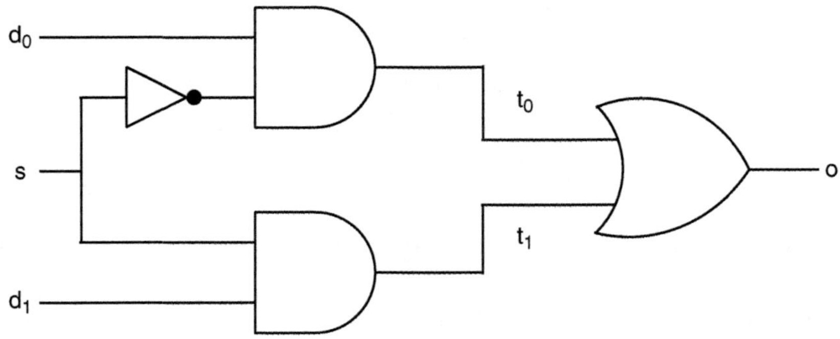

Figure 3.4: Example for suboptimal transformation

Now, consider the above transformation of the circuit into CNF. Under the partial assignment $d_0 = d_1 = 0$*, the value* 0 *will be implied at* o*. But under the partial assignment* $d_0 = d_1 = 1$ *standard BCP is not powerful enough to imply any further values since both AND gates, i.e. the auxiliary variables* t_0, t_1*, have a non-controlling value at one input. Thus, the value of the output cannot be implied.*

Now, assume that in the current state of the search process this partial assignment $d_0 = d_1 = 1$ *occurs and the next decision is* $o = 0$*. In this case, a conflict occurs immediately:* $o = 0$ *implies that the outputs of both AND gates are zero. This in turn causes contradictory assignments to* s*. The conflict clause* $(o + \overline{d}_0 + \overline{d}_1)$ *is created.*

An alternative procedure to transform the circuit could recognize the multiplexer structure and add the clause $(o + \overline{d}_0 + \overline{d}_1)$ *in advance to avoid backtracking.*

Of course, there is a trade-off between a powerful preprocessing step to simplify the CNF formula and the run time needed for this procedure. As explained above, SAT solvers often do some light-weight simplification on the clauses before starting the DPLL procedure.

Alternatively, the CNF representation can be simplified while transforming the circuit. The advantage is additional information about connectivity and functionality that cannot be recovered easily from the CNF. An AND gate with n inputs is a simple example. This gate may either be split into 2-input gates with one new variable per gate and three clauses per gate. While no additional variables are needed to directly represent the n-input AND gate with $n + 1$ clauses. If an XOR gate with n inputs is considered, the two level representation is exponential in the number of inputs. Thus, splitting the gate into multiple smaller gates is typically mandatory.

Further improvements can be obtained by dedicated data structures. So called AND-inverter graphs [64] directly simplify the circuit by partial canonization. The subsequent transformation of the graph into CNF allows for further simplification steps, e.g. [84]. In all these cases, the trade-off between run time and the benefit in solving the problem instance is crucial. Moreover, in the ATPG domain, a powerful preprocessing step that can be reused for all faults is useful. Such techniques will be further discussed throughout this book.

Nonetheless, a disadvantage of SAT-based ATPG is the loss of structural information. Only a set of clauses is presented to the SAT solver. Information about predecessors and successors of a node is lost and is not used during the SAT search. But as will be shown in the next chapter, this information can be partially recovered by introducing additional constraints into the SAT instance.

3.4 Circuit-Oriented SAT

As described in the previous section, to apply SAT solvers to circuit-oriented problems, the problem has to be transformed into a CNF formula. During this transformation, structural knowledge such as the connectivity is lost. Therefore, circuit-based SAT solvers and hybrid SAT solvers were proposed that retain structural information to speed up the search. Circuit-based SAT solvers work on a circuit structure, whereas hybrid SAT solvers work on CNF as well as on circuit structure.

In [53], a hybrid SAT technique is presented that uses the gate connectivity information in the inner loop of a branch-and-bound SAT algorithm. The connectivity information is extracted during the CNF transformation. Don't cares can be identified from this information and exploited to deactivate clauses.

The approach presented in [45] works partly on the circuit structure and partly on the CNF to combine the strength of the different representations. The original logical formula is processed in the circuit representation. In contrast, the conflict clauses are processed in CNF representation. Heuristics based on the circuit structure can be applied to speed up the search, whereas the conflict clauses can be processed by fast CNF-based BCP techniques.

In contrast, the circuit-based SAT solver CSAT [74,75] works completely on the circuit structure. Here, conflict clauses are represented as added gates. The implementation of CSAT is strongly influenced by the CNF-based SAT solver zChaff [82], but the SAT techniques are transferred to the circuit representation. As an advantage, structural information can easily be exploited.

CSAT strongly benefits from signal correlation guided learning. Signal correlations are identified during a simulation-based preprocessing and applied in two different ways: implicit and explicit.

In implicit learning, signal correlations are used to influence the decision strategy, whereas in explicit learning a correlated pair of signals is used to generate conflict clauses (recorded as gates). Such a pair of signals is assigned in a way that will most likely cause conflicts. The reasoning engine is then used to record conflict clauses (in the form of gates) derived from these assignments. The correlated pairs of signals are processed in topological order to reuse previously learned information. All learned information can then be utilized in the original problem instance. CSAT works especially well for problems with many signal correlations, e.g. equivalence checking problems.

In contrast to CSAT, the sequential SAT engine SATORI [58] carries out BCP for the original clauses as well as for the conflict clauses completely on CNF. The circuit structure is used for efficient decision heuristics motivated by ATPG algorithms to guide the search. Structural information is used to reduce the assignments to primary inputs and state variables. Illegal states are recorded in the form of conflict clauses.

The approach presented in [87] utilizes knowledge about the circuit structure to exploit observability don't cares statically as well as dynamically. In the first case, additional (redundant) clauses are added to the CNF in a preprocessing step. These clauses encode don't cares and indicate search space without any solution. Don't cares are applied dynamically by influencing the decision strategy. Using information about the circuit structure, don't cares are identified during the search process. Signals with don't care values that do not influence the outcome of the search are ignored in the decision heuristic and in the break condition.

The use of don't care information is improved in [40], where it is integrated into the basic SAT techniques. Certain don't care literals are added to clauses in the CNF representation during CNF transformation. These literals are used in conflict analysis to propagate the information during the learning process. The resulting conflict clauses also contain don't care information.

Recently, the circuit-based SAT solver QuteSAT was introduced in [15]. QuteSAT is similar to the circuit-based SAT solver CSAT but proposes a novel generic watching scheme on simple and complex gate types for BCP. Furthermore, it uses a new implicit implication graph to improve conflict-driven learning in a circuit-based SAT solver. QuteSAT focuses on efficient BCP techniques and data structures and does not contain structural techniques like implicit and explicit learning.

Chapter 4

SAT-Based ATPG

In this chapter,[1] SAT-based ATPG is explained in detail for the static SAFM. Dynamic fault models require a different modeling since two test vectors are required per fault. Path delay and transition delay faults are considered in Chapter 10.

The basic transformation of an ATPG problem to a SAT problem for the SAFM is presented in Section 4.1. This technique has been presented in [67] – but SAT solvers at that time were not very powerful. In particular, conflict-based learning was not available. Simply replacing the underlying SAT solver already provides substantial speed-ups in run time.

An improvement of the basic transformation by exploiting structural information, that has been proposed in [95], is shown in Section 4.2. Such problem specific improvements are a typical ingredient when applying a SAT solver to a known structured problem. Without encoding problem specific knowledge in the CNF formula, the SAT solver often spends a lot of time in learning information that is trivial when the higher level structure is known. Therefore, directly encoding such problem specific knowledge improves the overall performance of the SAT-based approach by enhancing the implicative power of the CNF representation. This in turn prevents wrong decisions and time consuming backtracking steps. Further enhancements and requirements for industrial application of these techniques are considered in subsequent chapters.

Section 4.3 provides experimental results for the introduced techniques. The proposed approach is compared to the SAT-based approach presented in [95] as well as against a classical structure-based algorithm. A summary of this chapter is given in Section 4.4.

---

[1]Parts of this chapter have been published in [26, 91].

R. Drechsler et al., *Test Pattern Generation using Boolean Proof Engines*,

4.1 Basic Problem Transformation

The transformation is formulated for a single fault. Then, the process iterates over all faults to generate a complete test set. Alternatively, the SAT-based engine can be integrated within an ATPG tool that uses other engines as well. This is considered in Chapter 9.

First, the fault site is located in the circuit. Then, the parts of the circuit that are structurally influenced by the fault site are calculated, as shown in Figure 4.1, by a depth first search.

All gates in the transitive fanout from the fault site towards the outputs are influenced, this is also called the *output cone*. Then, the fanin cone of all outputs in the output cone is calculated. These gates have to be considered when creating the ATPG instance for the particular fault.

Analogous to the construction of the Boolean difference shown in Figure 4.2, a fault free model and a faulty model of the circuit are joined into a single SAT instance to calculate the Boolean difference between both versions. This construction is similar to a *miter* circuit as introduced for equivalence checking [9], but for the case of ATPG, some parts that are known to be identical can be shared between the two copies of the circuit. All gates not contained in the transitive fanout of the fault site have the same behavior in both versions. Therefore, only the output cone is duplicated as shown in Figure 4.3.

Next, the Boolean differences of the corresponding outputs are calculated. At least, one of these Boolean differences must assume the value 1 to determine a test pattern for the fault. This corresponds to constraining the output of the OR gate in Figure 4.3 to the value 1. At this point, the complete ATPG instance has been formulated in terms of a circuit and can be transformed into CNF representation.

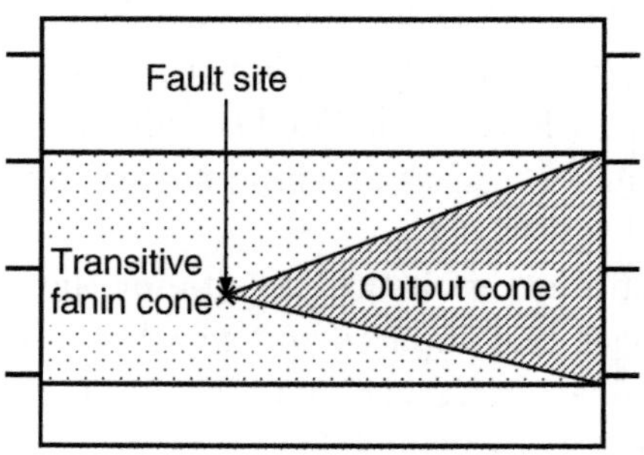

Figure 4.1: Influenced circuit parts

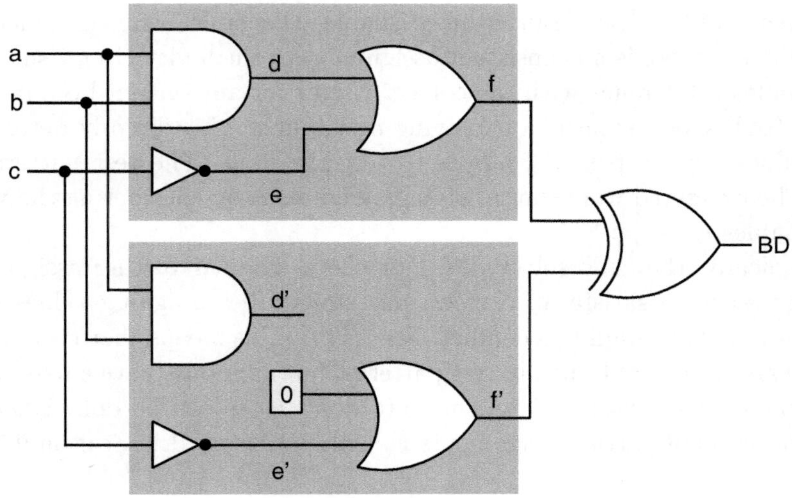

Figure 4.2: Boolean difference of faulty circuit and fault free circuit (replicated from Figure 2.3)

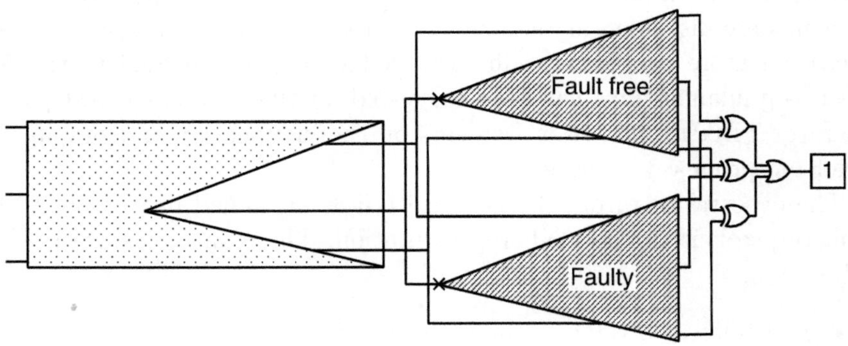

Figure 4.3: SAT instance for ATPG

Assigning values to those variables that correspond to the primary inputs $i_1, \ldots, i_n$ is sufficient. Once these assignments are done, BCP will imply the values of the internal gates and outputs. However, the SAT solver usually does not have this information in advance. Decisions may be made for internal gates as well. But, when the SAT solver finally returns a satisfying assignment, the values of primary inputs must be consistent with those of internal gates, since a combinational circuit is considered.

If the SAT solver returns unsatisfiable, the fault being considered is untestable – there is no consistent assignment to the faulty circuit such that the Boolean difference with the correct circuit for any output becomes one. If the SAT solver returns a satisfying assignment, this directly determines the values for the primary inputs to test the fault. The test pattern can easily be extracted from the satisfying assignment by considering the values of variables $i_1, \ldots, i_n$.

Typically, the SAT solver does not check whether further assignments are necessary to satisfy all clauses, but stops after assigning values to all variables without finding a conflict. For ATPG, this means that there are no don't care values contained in test patterns, but all inputs have a fixed value. Turning some of these assignments into don't cares can be done efficiently leading to small partial assignments as will be discussed in Section 9.3.

4.2 Structural Information

As explained earlier, most of the structural information is lost during the transformation of the original problem into CNF. This can be recovered by additional constraints. Generic ways to add clauses for any circuit-based problem were discussed in Section 3.3. In the particular application studied here, even more problem-specific knowledge can be provided to the SAT solver as guidance. This has been suggested for the SAT-based test pattern generator TEGUS [95]. Improvements on the justification and propagation have been proposed in [96, 98].

The observations from the D-algorithm as explained in Section 2.4 can be made explicit in the CNF representation. Three variables are used for each gate g:

- g_f denotes the value in the faulty circuit.
- g_c denotes the value in the correct circuit.
- g_d denotes, whether g is on a D-chain.

This notation supports the introduction of additional implications into the CNF:

- If g is on a D-chain, the values in the faulty and the correct circuit are different: $g_d \rightarrow (g_f \neq g_c)$.
- If g is on a D-chain, at least one successor of g must be on the D-chain as well: Let $h^i, 1 \leq i \leq q$ be the successors of g, then $g_d \rightarrow \bigvee_{i=1}^{q} h_d^i$.

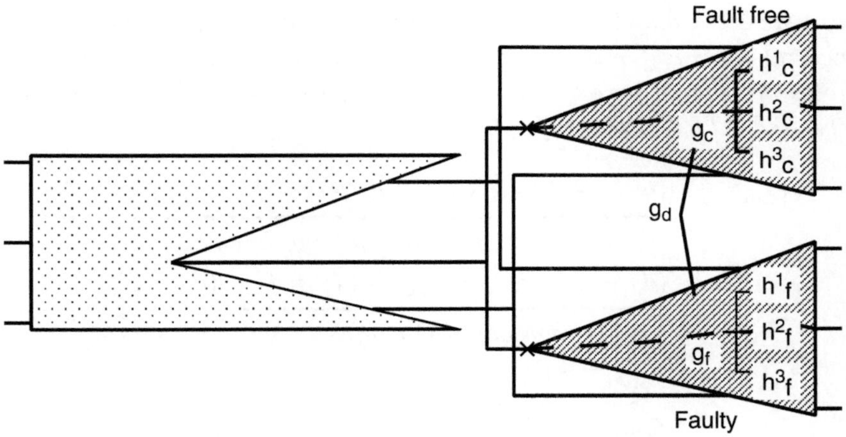

Figure 4.4: Embedding structural information

Figure 4.4 shows the beneficial effects of these additional clauses that simply seem to be an overhead on first consideration. Without the additional implications, the fault's output cone in the fault free version and the faulty version of the circuit are only connected via the variables on the cut to the shared portions of the circuit. In contrast, the g_d variables establish direct links between these structures.

As a result, implications are possible even when some variables in the shared fanin are not assigned yet. Moreover, the information about successors of a gate and the notion of D-chains are directly encoded in the SAT instance. Thus, the propagation of the fault observation along a certain gate immediately triggers the propagation along one of the successors. This goes on until a primary output is reached and a path from the fault site to the primary output is sensitized.

As a side effect, no logic to compare the output values of the fault free and the faulty circuit is necessary. The additional implications force the propagation towards at least one observable primary output. Those primary outputs where the fault effect can be observed are easily identified by considering the values of the variables $o_d^i, 1 \leq i \leq m$ that decide whether a primary output is part of a D-chain.

Since the constraint for D-chains is formulated as an implication, additional primary outputs may be used to observe the fault effect. Such outputs can be identified by comparing o_f^i to o_c^i. The test pattern is extracted from the satisfying assignment as before – the values of the variables $i_1, \ldots, i_n$ corresponding to primary inputs directly provide a test pattern.

Example 11 *Consider the example circuit in Figure 4.5. The circuit with structural sharing for the s-a-0 at d is shown in Figure 4.6. The gray box denotes the shared part of the problem instance. The following formula describes the ATPG problem for this fault:*

$$\Phi_{(d,0)} = \underbrace{(a \cdot b \leftrightarrow d_c) \cdot (c \leftrightarrow \overline{e}) \cdot (d_c + e \leftrightarrow f_c) \cdot (d_f + e \leftrightarrow f_f)}_{\text{constraints for the circuit}} \cdot \underbrace{(d_d \rightarrow (d_f \neq d_c)) \cdot (f_d \rightarrow (f_f \neq f_c))}_{\text{embedding D-chains}} \cdot \underbrace{(d_d \rightarrow f_d)}_{\text{structure}} \cdot \underbrace{(\overline{d}_f) \cdot (d_d)}_{\text{fault modeling}}$$

In this simple case, BCP iteratively does the following assignments (the actual order depends on the order of clauses in the watching lists, see Section 3.2.3):

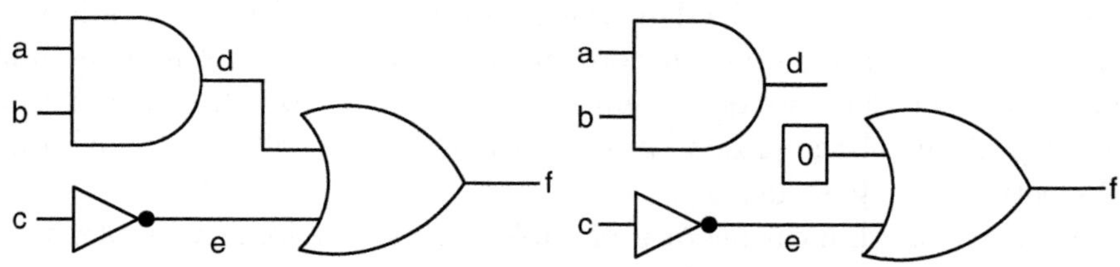

(a) Correct circuit (b) Faulty circuit

Figure 4.5: Example for the SAFM (replicated from Figure 2.2)

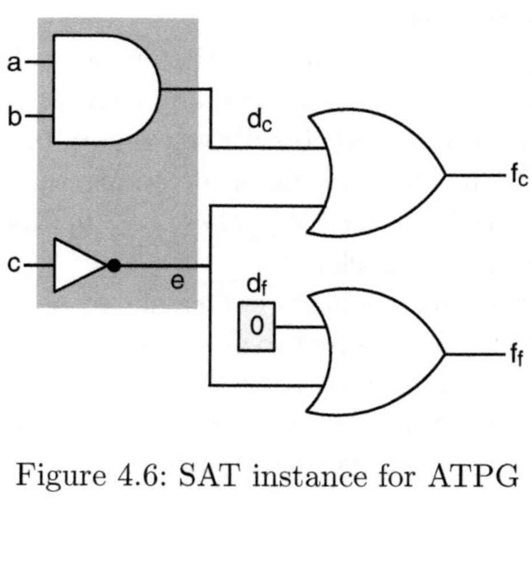

Figure 4.6: SAT instance for ATPG

$$d_c = 1, f_c = 1, f_d = 1, f_f = 0, e = 0, c = 1, a = 1, b = 1.$$

Here, no backtracking steps are necessary to generate the test pattern.

This underlying structure, i.e. the encoding of D-chains, is used throughout this book. Improvements are presented to generalize learned information, to handle large industrial circuits that may have non-Boolean elements and to improve the performance which is necessary for very large circuits.

A different model for ATPG is necessary when dynamic faults are considered – as two time steps are relevant in this case. Nonetheless, most of the improvements generalize to these cases as well. SAT instances for delay faults are discussed in Chapter 10.

4.3 Experimental Results

In this section, we report experimental results to show the improvements when advanced SAT techniques are applied. All experiments were carried out on an AMD Athlon XP 2200+ (1.8 GHz, 512 MByte RAM, GNU/Linux).

At this stage, PASSAT uses the SAT solver zChaff [82] that provides the advanced SAT techniques discussed in Chapter 3. PASSAT is compared to previous FAN-based and SAT-based approaches for ATPG. Instead of the industrial circuits, only the smaller ISCAS benchmark circuits are considered, as the previous tools reach their limits for these circuits.

Redundancy identification has been studied in the first experiment. As a hard example, a 40 bit multiplier has been considered. The results for TEGUS in comparison to PASSAT are shown in Figure 4.7. Run times (in CPU seconds) of an ATPG run for different redundant faults are reported. For a single problem instance, the run time and number of conflict clauses learned by PASSAT are determined, then the run time of TEGUS – without learning – is determined for the same instance. As can be seen, the run time of TEGUS grows significantly as the number of conflict clauses learned by PASSAT increases. PASSAT shows only a slight increase in run time. The redundancy is detected due to the unsatisfiability of the CNF formula.

TEGUS does not make use of conflict-based learning. Therefore, TEGUS exhausts the whole search space by checking all possible assignments before classifying the fault as redundant. In contrast, PASSAT prunes large parts of the search space due to conflict analysis and non-chronological backtracking, i.e by adding conflict clauses to the problem instance.

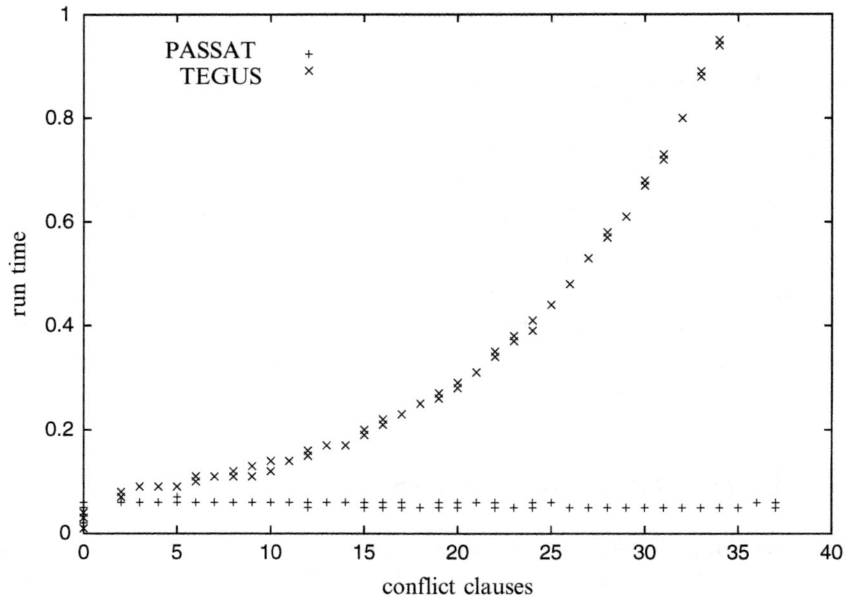

Figure 4.7: Redundancies: conflict clauses vs. run time

In the next series of experiments, the run time behavior of the two SAT-based approaches for the benchmarks from the ISCAS'85 and ISCAS'89 benchmark suites are studied. To demonstrate the quality of the SAT-based approaches, a comparison to an improved version of Atalanta [69], that is based on the FAN algorithm [44], is given. Atalanta was also used to generate test patterns for each fault. The backtrack limit was set to 10. The results are shown in Table 4.1.

The first column gives the name of the benchmark. Then, the run time is given in CPU seconds for each approach. Run times for Atalanta with and without fault simulation are given in Columns *Fs* and *No fs*, respectively. On circuits s35932 and s38584.1 Atalanta returned no results, when fault simulation was disabled. For TEGUS and PASSAT, the run time to generate the SAT instance and the time for SAT-solving are separately given in Columns *Eqn* and *SAT*, respectively.

Both SAT approaches are significantly faster than the classical FAN-based algorithm and solve all benchmarks in nearly no time. No fault simulation has been applied in the SAT approaches. Therefore, the run times should be compared to Atalanta without fault simulation. Especially for large circuits, the SAT approaches show a run time improvement of several

Table 4.1: Results for the Boolean circuit model

	Atalanta		TEGUS		PASSAT	
Circuit	Fs	No fs	Eqn	SAT	Eqn	SAT
c432	0.02	0.05	0.02	0.02	0.18	0.08
c880	0.03	0.13	0.00	0.02	0.16	0.02
c1355	0.02	0.90	0.05	0.02	0.29	0.21
c1908	0.12	1.00	0.06	0.00	0.54	0.10
c2670	0.58	3.12	0.04	0.08	0.79	0.12
c3540	0.85	5.73	0.23	0.16	3.17	0.66
c5315	1.12	17.70	0.08	0.04	1.17	0.27
c6288	0.75	49.43	0.21	0.10	4.78	1.79
c7552	5.72	65.93	0.20	0.42	2.61	0.70
s1494	0.08	0.37	0.01	0.00	0.06	0.01
s5378	1.70	18.37	0.03	0.02	0.37	0.06
s9234.1	18.63	83.90	0.14	0.39	3.06	0.47
s13207.1	18.63	127.40	0.29	0.16	3.03	0.61
s15850.1	27.13	204.27	0.68	0.76	7.66	1.52
s35932	87.40	*Timeout*	0.47	0.09	2.68	0.28
s38417	131.77	1624.78	0.52	0.24	3.56	0.65
s38584.1	86.30	*Timeout*	0.63	0.14	4.09	0.75

orders of magnitude. Even when fault simulation is enabled in Atalanta the SAT-based approaches are faster by up to more than two orders of magnitude (see e.g. s35932 and s38584.1).

Considering the SAT-based approaches, TEGUS is faster for these simple cases. Here, test patterns for all faults are generated. In this scenario, the speed-up gained by PASSAT for difficult untestable faults is overruled by the overhead for sophisticated variable selection and conflict analysis.

This shows that a careful integration with previous techniques is required to use a very powerful highly optimized algorithm in an environment where simpler algorithms with lower overhead are often sufficient. Such an integration will be presented in Chapter 9.

4.4 Summary

This chapter introduced the basic model for SAT-based ATPG. Improvements have been presented that encode structural information into the SAT instance. The experiments show a significant speed-up of the SAT-based techniques over a public domain implementation of the FAN algorithm.

Moreover, advanced SAT techniques yield a speed-up in comparison to a simple backtrack search when hard untestable instances of test pattern generation are considered.

Using SAT-based ATPG for large industrial circuits requires further effort in modeling, instance generation and integration with existing frameworks. These issues are studied throughout the next chapters.

Chapter 5

Learning Techniques

Intuitively, the ATPG problem is appropriate for learning strategies. The same underlying structure – a circuit – is iteratively considered for a large number of very similar problems, i.e. generating test vectors for all faults. Thus, learning techniques have traditionally been exploited in ATPG.

SOCRATES [88] was one of the first tools to incorporate learning into the standard ATPG flow. Implications on reconvergent circuit structures were statically added during a preprocessing step, but only certain types of implications were detected. Subsequently, the more general concept of recursive learning [66] was applied in the ATPG tool HANNIBAL. Recursive learning is complete, in principle, i.e. all implications resulting from a partial assignment of values to signals in a circuit can be learned. However finding the cause for a certain implication on the circuit structure requires too many backtracking steps in practice. Therefore, this technique is not feasible for large circuits.

In traditional ATPG tools learning typically means an overhead that has to pay off while generating test patterns. In contrast, the SAT solver at the core of a SAT-based ATPG tool uses learning in the core algorithm to traverse the search space more effectively.

As explained in the previous chapters, a SAT solver learns information from a SAT instance each time a non-solution subspace is found [79]. Conflict clauses serve as an efficient data structure to store this information. Techniques to reuse learned information for similar SAT instances have been proposed [107] and applied e.g. for bounded model checking [52, 92]. The challenge for reuse lies in the creation of the SAT instance and storing the learned information in a database. Domain specific knowledge is needed to allow for efficient reuse of this information.

R. Drechsler et al., *Test Pattern Generation using Boolean Proof Engines*,

Learning results for SAT-based ATPG have been reported for stuck-at faults [77, 78]. The approach is similar to the techniques considered here. Learning for path delay faults has been considered in [16], but in this case dynamic learning is based on the time consuming calculation of unsatisfiable cores.

In this chapter,[1] two strategies to reuse dynamically learned information for SAT-based ATPG of stuck-at faults are considered. The first approach makes use of *incremental SAT* [107]. In this paradigm, the SAT solver is never released, i.e. not reinitialized between faults, but the SAT instance is modified on the fly. So learned information is kept if applicable. A heuristic to enumerate stuck-at faults such that subsequent SAT instances are very similar is proposed.

In the second approach, a more general circuit-based learning scheme is applied. This is necessary when SAT-based ATPG is applied in a multi-engine environment as it is usually done in industrial practice. The correctness of this learning approach is proven. Both techniques are applied to publicly available benchmark circuits and large industrial circuits. The experimental results show that the performance and robustness of SAT-based ATPG are significantly improved.

This chapter is structured as follows. A first introductory example is given in the following section. The basics of incremental SAT and reusing learned information in SAT solvers and implementation issues are briefly reviewed in Section 5.2. In Section 5.3, the heuristic to apply incremental SAT is presented. Then, the approach for circuit-based learning is introduced and proven to be correct. Experimental results for a range of benchmark circuits are given in Section 5.4. Conclusions are presented in the final section.

5.1 Introductory Example

Consider the circuit structure shown in Figure 5.1 which is assumed to be part of a larger circuit. As explained in Example 10 in Section 3.3, this is a multiplexer. When both data inputs have the value 1, the output takes the value 1, but this cannot be implied from the CNF unless the value of s has been assigned. Thus, the conflict clause $(\overline{d}_0 + \overline{d}_1 + o)$ may be learned during the search. Nonetheless, the implication $(d_0 \cdot d_1) \rightarrow o$ holds for this sub-circuit in general. Therefore, this implication (or clause) can always be reused when the circuit structure is contained in the ATPG instance.

[1]Parts of this chapter have been published in [37, 106].

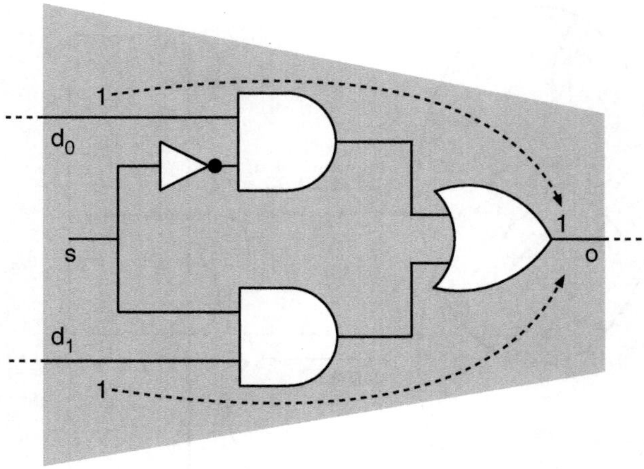

Figure 5.1: Example for reuse

Of course, this is a simple example that may be handled in a preprocessing step. But finding all circuit structures where the gate-wise traversal leads to a CNF representation where some implication cannot be inferred by BCP would be too time consuming and would require a huge database. Thus, learning such information from the circuit while the SAT solver is running and reusing the information as needed is a potential solution.

In both "worlds" – ATPG and SAT solving – the idea of learning has been known for quite some time. The concepts that evolved in the SAT community are briefly discussed next. Then, these ideas are transferred to and exploited for ATPG.

5.2 Concepts for Reusing Learned Information

5.2.1 Basic Idea

Incremental SAT has been proposed to reuse learned information when a series of structurally similar SAT instances has to be solved [107]. Consider two CNF formulas Φ_A, Φ_B that are given as sets of clauses where Φ_A is solved first. Then, all clauses learned from $\Phi_A \cap \Phi_B$ can directly be applied when solving Φ_B. Reusing the learned information can speed up the solving process for Φ_B.

Technically, a SAT solver learns in terms of conflict clauses (see Section 3.2). Logically, a conflict clause κ, learned while solving Φ_A, can be derived from Φ_A using resolution: $\Phi_A \models \kappa$. The resolution steps needed,

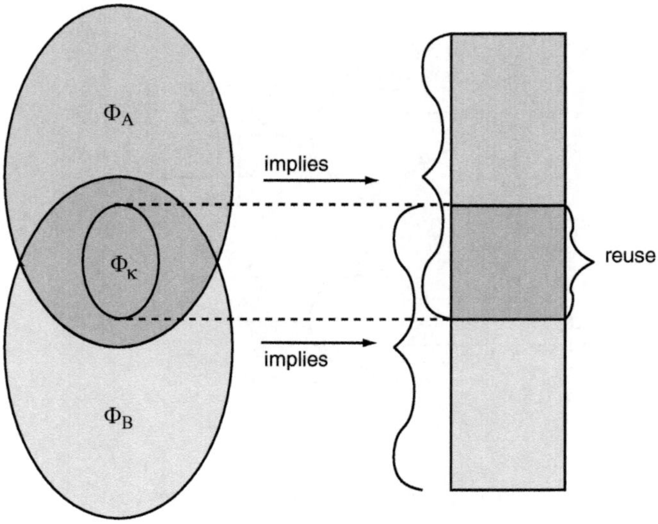

Figure 5.2: Learning from two SAT instances

i.e. which clauses have to be resolved, are stored in the implication graph of the SAT solver. Therefore, for any conflict clause κ, the subset of original clauses $\Phi_\kappa \subseteq \Phi_A$ that implied κ can be determined, i.e. $\Phi_\kappa \models \kappa$. Then, κ may be reused for Φ_B if $\Phi_\kappa \subseteq (\Phi_A \cap \Phi_B)$. This is illustrated in Figure 5.2.

To apply this approach in practice, some requirements have to be met:

- For a series of SAT instances, the intersection has to be known.
- Given a conflict clause κ, the subset Φ_κ has to be derived efficiently.

Both of these requirements offer flexibility for a trade-off between accuracy and efficiency. First, underapproximating the intersection between subsequent SAT instances is safe. Assume that instead of all common clauses between Φ_A and Φ_B, only a subset $I \subset (\Phi_A \cap \Phi_B)$ is used. Now, assume for a conflict clause κ that $(\Phi_A \cap \Phi_B) \models \kappa$ and $I \not\models \kappa$. In this case, κ is excluded from reuse when solving Φ_B, even though the learned information is valid in Φ_B.

Nonetheless, such an underapproximation may be useful in identifying the intersecting sets more easily. How to effectively identify these sets strongly depends on the application and is discussed in Section 5.3 for SAT-based ATPG.

Similarly, a more accurate estimation of Φ_κ typically requires higher overhead during the SAT procedure. On the other hand, overapproximating

Φ_κ by identifying a larger set $\Phi_{\kappa'} \supset \Phi_\kappa$ is save. Whenever $\Phi_{\kappa'} \subseteq (\Phi_A \cap \Phi_B)$, clause κ can be reused safely, but some conflict clauses may be discarded, even though they could be reused.

5.2.2 Tracking Conflict Clauses

The implementation of the mechanism to track the "reasons" for conflicts happening is typically tightly integrated into the algorithms of the SAT solver. Three possible approaches are briefly discussed.

The first uses the propagation of "tags" as introduced in [92] in the context of bounded model checking [7]. Assume that it is known from the application which clauses are in $\Phi_A \cap \Phi_B$. A clause in $\Phi_A \cap \Phi_B$ is initially marked using a tag t:

$$t(\kappa) = \begin{cases} 1, \text{if } \kappa \in (\Phi_A \cap \Phi_B) \text{ (marked)} \\ 0, \text{otherwise (unmarked)} \end{cases}$$

Then the SAT solver is called to find a solution for Φ_A. As explained in Section 3.2, the solver analyzes the implication graph to generate conflict clauses. The same mechanism is used to identify which clauses are necessary to derive a conflict clause. A conflict clause ω may be reused for Φ_B when all clauses that are necessary to derive ω are marked. In this case, the new conflict clause is also marked. If at least one clause is not marked, the conflict clause may not be reused.

Example 12 *Let $\Phi_A = \{\omega_1, \omega_2, \ldots, \omega_5\}$ and $\Phi_B = \{\omega_4, \omega_5, \ldots, \omega_8\}$. Thus, $t(\kappa) = 1$ for $\kappa \in \{\omega_1, \omega_2, \omega_3\}$. Now, assume while solving Φ_A, the situation denoted by the conflict graph shown in Figure 5.3a occurs and the conflict clause $\omega_\kappa = \{a, \overline{\kappa}\}$ is derived. Only clauses ω_1 and ω_2 are involved in the conflict. Thus, ω_κ may be reused and is also marked.*

Alternatively, consider the conflict shown in Figure 5.3b. In this case, clause ω_6 also participates in the conflict, but is not marked. Thus, the new conflict clause cannot be reused and is therefore not marked.

The procedure available in the SAT solver zChaff [8,82] is an extension of the above technique allowing the tracking of the origins of conflict clauses more accurately. In zChaff, each original clause κ can be associated to a group $G(\kappa)$. Groups of clauses can be added, deleted or extended without recreating the whole SAT instance.

The information about groups for a clause κ is a bit vector $G(\kappa)$. For the initial clauses, typically exactly one bit is set to 1, while all others

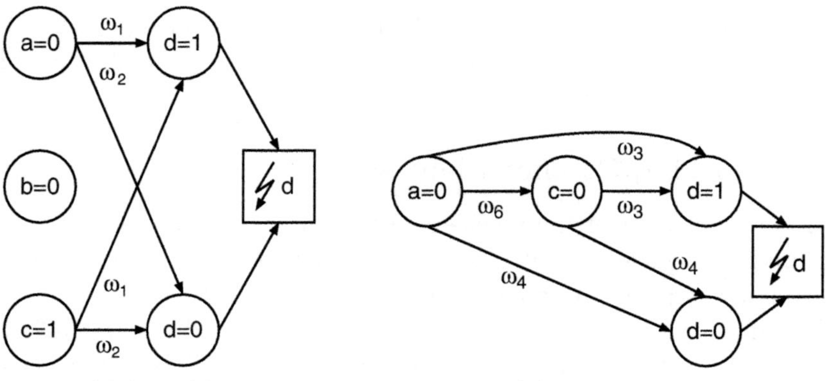

(a) Reusable (b) Non-reusable

Figure 5.3: Conflicts while solving Φ_A

are 0. Hence, the initial clauses are associated to exactly one group. Now, let the conflict clause ω_κ be derived from clauses $\omega_1, \ldots, \omega_n$. Then, the group $G(\omega_\kappa)$ is defined as the bit-wise OR of the bit-vector tags of the participating clauses, i.e. using the corresponding operator of the C programming language

$$G(\omega_\kappa) = |_{i=1}^{n} G(\omega_i).$$

Thus, each clause may belong to multiple groups. Whenever a group is deleted, all clauses belonging to this group are deleted as well. Or alternatively, the groups necessary to derive a certain conflict clause are described by the associated bit vector. In zChaff, this bit vector is chosen to match the width of a data word on the machine (i.e. typically 64 bit) to keep the overhead moderate.

Finally, an even more accurate approach has been proposed in [90]. The target application there was the construction of small unsatisfiable cores for unsatisfiable SAT instances, i.e. to derive a subset of clauses that is unsatisfiable already (see Section 3.2.4). In this case, all clauses responsible for the final conflict have to be identified. The underlying idea is the same: by traversing the implication graph, reasons for conflicts are determined and stored.

However, keeping the complete implication graph is too expensive. Thus, the graph is pruned as soon as decisions are undone; only the resulting conflict clauses are kept. The approach presented in [90] suggests adding a reference counting scheme to manage the implication graph.

Conflict nodes in the graph are referenced by conflict clauses. As long as a certain conflict clause is not deleted, the corresponding conflict node and all preceding nodes in the graph are kept. When a conflict clause is deleted, the conflict node is dereferenced and may be deleted and the reference counts for the supporting clauses are decremented. A conflict clause may only be deleted if it does not participate in any further conflicts, i.e. its reference count becomes 0.

Obviously, this mechanism accurately reproduces all steps necessary to derive the final conflict for an unsatisfiable SAT instance, i.e. an empty clause. The experiments showed that the overhead in terms of memory and run time is still acceptable.

The grouping scheme of zChaff is used in the approach described below. For SAT-based ATPG, groups of clauses can be identified quite efficiently. Two scenarios are considered. First, the solver is not released after solving an instance. Instead groups of clauses are modified on the fly. Thus, the interface for groups of clauses entirely manages the deletion and reuse of conflict clauses for subsequent SAT instances. Second, another abstraction level is applied by collecting clauses in an external database. Preconditions that have to be fulfilled for reuse are derived for such stored conflict clauses.

5.3 Heuristics for ATPG

5.3.1 Notation

The instance for SAT-based ATPG has to be split into several groups or subsets to reuse conflict clauses. Given a circuit and a stuck-at fault A, the SAT instance Φ_A is created that is only satisfiable if a test pattern for A exists. If Φ_A is unsatisfiable the fault is redundant. Figure 5.4 shows the structure of Φ_A (this is a refinement of Figure 4.3 in Section 4.1). Essentially, Φ_A contains a model of the circuit without the fault and a model with the fault. Then, the SAT solver searches for an input assignment that forces at least one output to different values in the two models. Parts of the models are shared.

Constraints to model different parts of the SAT instance are denoted as follows:

- Ω_A^{corr} – the gates in the correct model of the output cone (i.e. the fanout cone of the fault site) or the shared part of the circuit
- Ω_A^{faulty} – the faulty part of the circuit
- Ω_A^{diff} – forces a difference at at least one output
- Ω'_A – the faulty copy of the gate at A

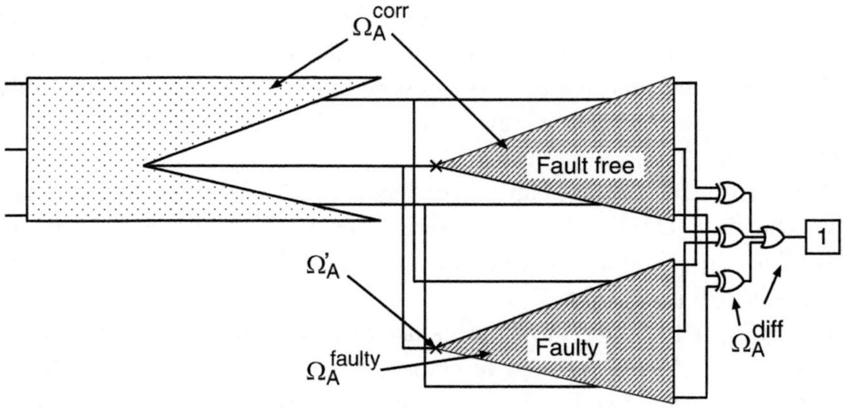

Figure 5.4: Structure of Φ_A

Thus,

$$\Phi_A = \Omega_A^{corr} \cup \Omega_A^{faulty} \cup \Omega_A^{diff} \cup \Omega'_A.$$

In the following, for a gate g, the constraints contained in Ω_A^{corr} are denoted by Ω_g.

This is only a simplistic illustration of the SAT instance for an ATPG problem. As discussed in Section 4.2 additional constraints, i.e. the encoding of D-chains, are added to the SAT instance in practice to reflect the structure of the circuit. This makes SAT solving more efficient. The extension of the proposed approach for reuse to these cases is straightforward. Essentially, the additional constraints can be handled as a separate group. For clarity of the explanation, these issues are not discussed further in this chapter.

5.3.2 Incremental SAT-Based ATPG

In the context of ATPG, the ordering of all faults determines the series of SAT instances considered. The objective is to order the faults such that subsequent SAT instances

1. Have large identical parts and

2. The identities can be determined efficiently

A heuristic to partition the set of faults has been developed. All faults in a single partition are handled incrementally. The clauses in the SAT instance are grouped depending on the heuristic. While enumerating faults in a single partition, some groups of clauses are kept while others are replaced. When continuing with the next fault partition, the whole SAT instance is rebuilt and all learned information is dropped.

As an extreme, each fault can be put into a separate partition. This corresponds to independent calls of the SAT solver for each fault. No information is reused. This partitioning is called *Classic* in the following.

On the other hand, all faults can be stored in a single partition. Then, the fault free part of the circuit always contains a model of the whole circuit. In this approach, clauses are never dropped. Learned information is accumulated during ATPG, but may cause a significant overhead in the size of the SAT instance. This partitioning is called *TotalInc.*

More promising is a compromise between these two extremes. In the following, the *Gate-input-partitioning* heuristic is described. A partition contains all faults at the inputs of a gate. An example for this partitioning is shown in Figure 5.5. Six partitions are created indicated by the gray boxes. Each 'x' denotes a stuck-at fault. At each gate, at most two stuck-at faults are possible, i.e. the s-a-0 and the s-a-1. Note that no fault collapsing is considered in the figure, but fault collapsing is applied as a preprocessing step in the experiments.

Given a gate g, large parts of the SAT instances that correspond to faults in a single partition are identical:

- Output cone: Due to the use of fanout gates, all fault locations have the same paths to the primary outputs. Therefore, the output cone is identical for all faults in the partition. This is valid for the faulty part and the fault free part of the circuit.

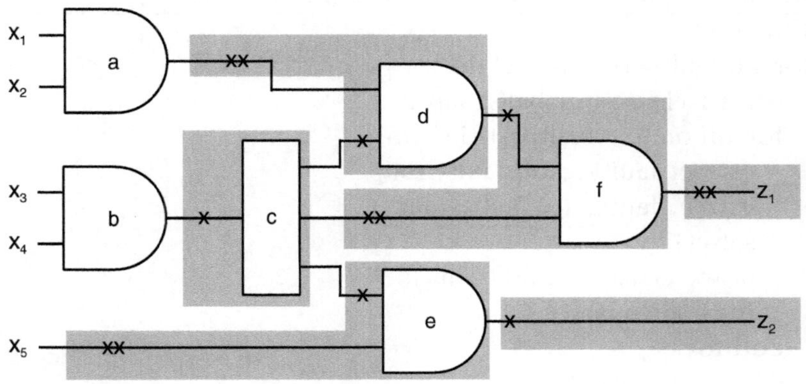

Figure 5.5: Example for gate-input-partitioning

- Fault free part of the circuit: The fault free part contained in the SAT instance is determined by traversing the circuit from outputs in the output cone of the fault towards the inputs. Because the output cones are identical, the fault free part of the circuit is also identical.

All clauses corresponding to these parts, i.e. $\Omega_A^{corr} \cup \Omega_A^{faulty}$, are summarized in the group *globalGroup* of clauses.

The only difference between two SAT instances is the model of the gate that is considered faulty. Different clauses are needed to model the stuck-at value at different inputs of the gate. Also, the two stuck-at faults at a single input differ in their value. Therefore, all clauses to model the gate and the fault value (Ω'_A) are collected in the group *faultGroup*. The overall ATPG procedure for gate-input-partitioning is shown in Algorithm 1. All partitions are enumerated. The function *extract_clauses(globalGroup)* creates the clauses in $\Omega_A^{corr} \cup \Omega_A^{faulty}$.

These clauses are stored in *globalGroup* and are not changed while enumerating other faults in the current partition. Then, all faults within the partition are handled individually. The clauses to encode Ω'_A, i.e. to model the faulty gate (*extract_faulty_gate*) and the fault value (*extract_fault_site*), are created and stored in *faultGroup*. By solving the SAT instance, the function *solve* classifies the fault as testable or untestable.

Afterwards, all clauses in *faultGroup* and all clauses derived from this group are removed by calling the procedure *delete_clauses*. Finally, to restart the search, the SAT solver has to be reset before proceeding to the next fault. Only when a new partition is considered, are all clauses removed.

Algorithm 1 Algorithm based on gate-input-partitioning

```
Circuit C;
for all faultpartition ∈ C do
  extract_clauses( globalGroup );
  for all fault ∈ faultpartition do
    extract_faulty_gate( faultGroup );
    extract_fault_site( faultGroup );
    solve();
    delete_clauses( faultGroup );
    reset_sat_solver();
  end for
  delete_all_clauses();
end for
```

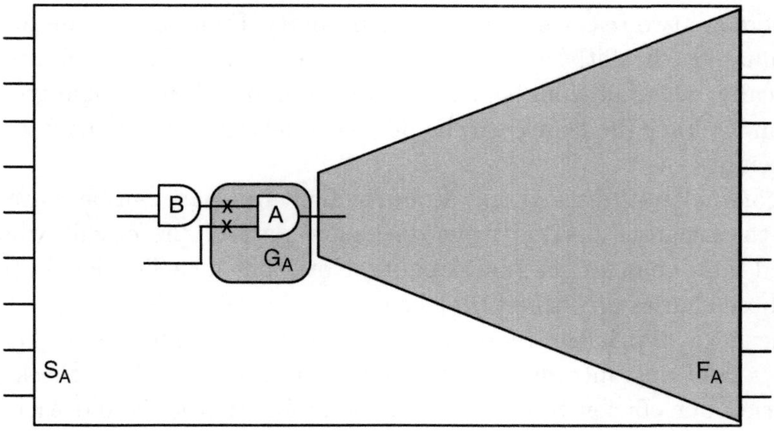

Figure 5.6: Problem instance for a single gate-input-partition

Other heuristics besides gate-input-partitioning have been implemented and evaluated, e.g. by grouping faults along paths, at outputs or a combination of these heuristics. Results for these heuristics are given in [106]. Gate-input-partitioning has been found to be the most efficient partitioning scheme in the experiments regarding run time and memory consumption. This partitioning scheme is depicted in Figure 5.6.

5.3.3 Enhanced Circuit-Based Learning

In practice, statically partitioning all faults during preprocessing is not feasible. Many faults are dropped from the fault list as a result of fault simulation. To be more efficient, learning should be circuit-based and should also be independent from the SAT instance and the SAT engine. In this section, an efficient circuit-based learning strategy is provided and the correctness of the approach is proven.

First, learned clauses are stored in a database, then stored clauses are considered for reuse. In the database, a learned clause κ is stored as a set of literals $\{\lambda_0, \ldots, \lambda_n\}$. A variable in the SAT instance corresponds to the output of a gate. Therefore, each literal λ_i is a pair (g_i, P_i) where g_i denotes the gate and P_i the polarity. $P_i = 0$ denotes the negative literal, $P_i = 1$ denotes the positive literal. After solving a SAT instance, the learned clauses are analyzed and stored in the database if they satisfy the following property:

Property: Clause κ is derived from the fault free part of the circuit, i.e. $\Omega_A^{corr} \models \kappa$.

There are two reasons to apply this property. First, the precondition can be evaluated efficiently across all SAT instances for different faults. More specifically, when all clauses in Ω_A^{corr} are summarized in a single group, the decision whether the clause can be derived solely from the fault free part of the circuit is easy.

Second, clauses derived only from the fault free part can be reused more easily than clauses derived from the faulty part of the circuit where the injected fault changes the functionality. Note that for efficiency, in practice only those clauses are stored that have three literals or less.

The next step is the reuse of stored clauses. Inserting a stored clause κ into a SAT instance Φ_A is only allowed if $\Phi_A \models \kappa$. This check has to be carried out efficiently because it is done for each fault and each stored clause. Such, an efficient check is provided and the correctness is proven below. In this context, it is sufficient to check whether Φ_A contains clauses for all gates that are considered in κ.

Before formally proving the soundness of this approach, Figure 5.7 illustrates the underlying idea. For a SAT instance Φ_A, only the correct part Ω_A^{corr} is shown which is essentially a portion of the correct circuit. All literals of a stored clause $\{\lambda_1, \lambda_2, \lambda_3\}$ correspond to certain signal lines in the circuit. Whenever a signal line is contained in Ω_A^{corr}, its complete fanin is also contained. All dependencies between signal lines can be derived by considering only their joined fanin – neither the fanout cone nor constraints for fault modeling or D-values are relevant. Thus, when all variables of a stored clause are contained in Ω_A^{corr}, then $\Phi_A \models \kappa$. The clause can be reused without changing the solution space. The following two lemmas help to prove the main result.

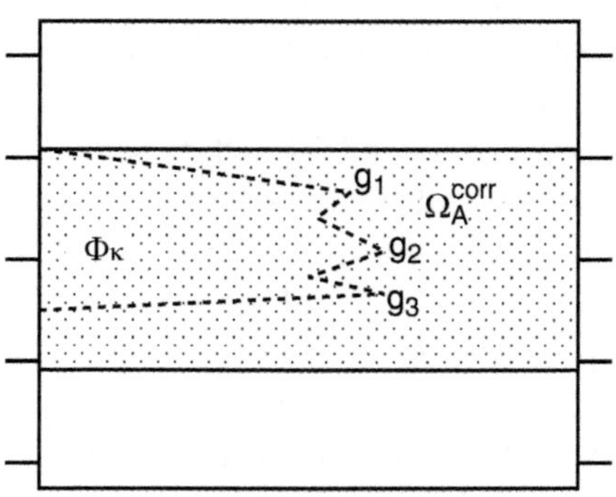

Figure 5.7: Idea of the proof

Lemma 1 *Let Φ_A be a SAT instance for stuck-at fault A and for gate g let $\Omega_g \subseteq \Phi_A$. Then, for any gate h in the transitive fanin $\mathcal{F}(g)$ of g, $\Omega_h \subseteq \Phi_A$.*

Proof. Due to construction $\Omega_g \subseteq \Omega_A^{corr} \subseteq \Phi_A$. Constraints for gate g are only inserted if g is reached while traversing the circuit towards the primary inputs. Then, constraints for all gates in $\mathcal{F}(g)$ are also inserted into Φ_A.

□

Lemma 2 *Let $\kappa = \{\lambda_1, \ldots, \lambda_n\}$ be a stored clause, Φ_A be a SAT instance for stuck-at fault A and $\Phi_A \models \kappa$. Let*

$$\mathbb{G} = \{g : (g, P) \in \kappa, \textit{where } g \textit{ is a gate and } P \in \{0, 1\}\}.$$

Then, κ can be implied by considering only

$$\Phi_\kappa = \bigcup_{h \in \mathcal{F}(\mathbb{G})} \Omega_h,$$

i.e. all clauses that correspond to gates in the fanin of $\mathbb{G}$.

Proof. According to the rule for storing clauses, it is sufficient to consider Ω_A^{corr}. Due to construction, $\Phi_\kappa \subseteq \Omega_A^{corr}$ holds.

Given the values of all but one gate, the value of the last gate can be implied. The clause κ corresponds to the Boolean expression $\lambda_1 + \ldots + \lambda_n$ that can be rewritten as $\overline{\lambda_1} \cdot \ldots \cdot \overline{\lambda_{n-1}} \to \lambda_n$ (without loss of generality any other literal than λ_n may be chosen, choosing n simplifies the notation).

The value of a gate g only depends on its predecessors in the circuit, i.e. on $\mathcal{F}(g)$. Let Ξ be a CNF that is only satisfied by an assignment to the primary inputs that forces all gates $g_i, i < n$ to the values $\overline{P_i}$.

First, assume no such assignment exists. Then, $\Phi_\kappa \to \kappa$ holds because the antecedent $\overline{\lambda_1} \cdot \ldots \cdot \overline{\lambda_{n-1}}$ is never satisfied.

Otherwise, such an assignment exists. Then, the CNF $\Phi_\kappa \cup \Xi$ can only be satisfied under a variable assignment if g_n assumes the value P_n because $\overline{\lambda_1} \cdot \ldots \cdot \overline{\lambda_{n-1}} \to \lambda_n$ holds on Ω_A^{corr}. Thus, $(\Phi_\kappa \cup \Xi) \to \lambda_n$ holds. By construction, the constraint Ξ is equivalent to the Boolean expression $\overline{\lambda_1} \cdot \ldots \cdot \overline{\lambda_{n-1}}$ or in set notation to the CNF $\{\{\overline{\lambda_1}\}, \ldots, \{\overline{\lambda_{n-1}}\}\}$ with respect to Φ_κ. Thus, $(\Phi_\kappa \cup \{\{\overline{\lambda_1}\}, \ldots, \{\overline{\lambda_{n-1}}\}\}) \to \lambda_n$. Therefore, if $\overline{\lambda_1} \cdot \ldots \cdot \overline{\lambda_{n-1}}$ is satisfied, Φ_κ can only be satisfied if λ_n is satisfied.

This leads to $\Phi_\kappa \models \kappa$.

□

Theorem 1 *Let* $\kappa = \{\lambda_1, \ldots, \lambda_n\}$ *be a stored clause and* Φ_A *be a SAT instance for stuck-at fault A. Further, for each* $i \in \{1, \ldots, n\}$ *and* $\lambda_i = (g_i, P_i)$ *let* $\omega_{g_i} \subseteq \Phi_A$. *Then,* $\Phi_A \models \kappa$.

Proof. Clause κ was learned previously on a SAT instance Φ_B for stuck-at fault B. According to Lemma 2, clause κ can be implied by Φ_κ (as defined in the lemma). Furthermore, $\Phi_\kappa \subseteq \Phi_A$ according to Lemma 1. Thus, $\Phi_A \models \kappa$.

□

Based on this foundation, two learning approaches have been proposed. First, we applied learning only in a preprocessing step. For each output, the circuit is converted into a CNF and the SAT solver is started on this CNF. The learned clauses of this run are considered for creating a *static* database. The second approach applies *dynamic* learning. After running the SAT solver on the SAT instance for a particular fault, the database is updated with the learned clauses.

5.4 Experimental Results

In the experiments, benchmark circuits from the ISCAS'85, ISCAS'89 and ITC'99 benchmark suites as well as industrial circuits from NXP Semiconductors are considered. Statistical information about the industrial circuits is provided in Section 2.5.

All experiments were carried out on an AMD Athlon XP 64 3500+ (2,2 GHz, 1,024 MByte RAM, GNU/Linux). The proposed learning techniques are implemented on top of the SAT-based ATPG tool PASSAT introduced in the previous chapter. Moreover, results for the application of PASSAT to the industrial circuits from NXP Semiconductors are reported to prove the benefit of learning in this context.

For this purpose, PASSAT applies a four-valued logic to handle circuits containing multiple-valued logic such as tri-state values and unknown values coming from the environment of the circuit. Details on the Boolean encoding of this multiple-valued logic are discussed in Chapter 6. The SAT solver zChaff [82] in the 2004 version which provides an interface for incremental SAT was used.

For each circuit, all stuck-at faults are classified using the SAT-based engine. No other engines and no fault simulation are applied (which can further speed up ATPG in practice). Fault collapsing is used to reduce the number of faults in advance. For each remaining fault a time out of 20 CPU seconds was applied. Additionally, the proposed learning techniques were embedded.

Table 5.1: Run time for incremental SAT

Circuit	Classic		Gate-input			TotalInc	
	Eqn	SAT	Eqn	SAT	Imp.	Eqn	SAT
c432	3.0	1.4	1.3	1.3	1.69	6.3	6.1
c499	10.0	54.6	4.7	35.0	1.63	30.5	61.0
c1355	17.4	83.7	6.6	43.5	2.02	45.7	86.1
c1908	13.2	15.9	5.8	12.5	1.59	45.6	51.7
c3540	49.4	37.7	20.2	31.4	1.69	167.5	157.0
c7552	102.2	130.6	46.7	93.3	1.66	449.5	536.3
s1494	2.1	1.7	1.0	1.7	1.41	8.4	10.1
s5378	19.5	7.6	8.7	5.5	1.91	111.9	132.7
s15850	145.6	70.9	66.8	58.6	1.73	1,693.5	1,318.7
s38417	220.0	88.1	95.8	70.8	1.85	*Mem. out*	
b10_C	0.5	0.2	0.2	0.1	2.33	1.2	1.0
b11_C	6.4	2.2	2.8	1.8	1.87	19.6	20.8
b12_C	6.8	3.3	2.8	2.7	1.84	47.8	51.6
b14_C	856.9	2,485.1	391.7	1,921.2	1.44	*Mem. out*	
b15_C	1,310.9	4,511.9	555.0	3,432.5	1.46	*Mem. out*	
				Avg.	1.74		

Results for the application of incremental SAT are shown in Table 5.1. Data is presented for the partitionings *Classic*, *Gate-input* and *TotalInc* as explained in Section 5.3.2. For each algorithm, the total run times for generating the SAT instances (*Eqn*) and solving (*SAT*) are reported in CPU seconds. The speed-up of gate-input-partitioning vs. classic is also reported (*Imp*). Even *Classic* classified all faults within the time limit, i.e. no aborts occurred.

Compared to the classical approach gate-input-partitioning provides remarkable speed-ups. The generation of the SAT instances is done much faster because large parts are simply reused. Also, the time for solving the problems is significantly reduced due to the learned clauses. On average, a speed-up of 1.74 was obtained on the benchmarks. The memory needs for gate-input-partitioning were the same as for the algorithm *Classic*.

In contrast, *TotalInc* causes a drastic increase in memory use due to a large number of learned clauses that were accumulated while enumerating all faults. As a result, the run time increased and in some cases the memory limit of 1,250 MByte (including swapping space) was exceeded.

Table 5.2: Run time of learning on top of gate-input-partitioning

Circuit	Gate-inp.	Static		Dynamic	
	Time	Time	Imp.	Time	Imp.
c432	2.6	2.7	0.96	2.6	1.00
c499	39.7	30.7	1.29	21.0	1.89
c1355	50.1	40.0	1.25	32.5	1.54
c1908	18.3	16.9	1.08	14.4	1.27
c3540	51.6	54.1	0.95	47.9	1.07
c7552	140.1	145.6	0.96	106.5	1.31
s1494	2.7	2.7	1.00	2.8	0.96
s5378	14.2	15.5	0.91	14.3	0.99
s15850	124.4	139.3	0.89	121.3	1.02
s38417	166.6	191.3	0.87	226.0	0.73
b10_C	0.3	0.4	0.75	0.3	1.00
b11_C	4.6	4.8	0.95	5.1	0.90
b12_C	5.5	5.6	0.98	5.6	0.98
b14_C	2,312.9	1,982.6	1.16	1,426.8	1.62
b15_C	3,987.5	3,665.3	1.08	2,673.6	1.49
		Avg.	1.00	Avg.	1.18

Next, the two circuit-based learning approaches are applied to the algorithm based on gate-input-partitioning. Experimental results for the combination with gate-input-partitioning are reported in Table 5.2. Here, the improvements are reported in comparison to gate-input-partitioning without learning.

When gate-input-partitioning is used, the preprocessing does not improve the overall performance. The learned clauses stem from "simple" conflicts and do not improve the performance for hard SAT instances. In contrast, the dynamic approach that analyzes and stores learned clauses after each run of the SAT solver improves the performance on average by another 18% over gate-input-partitioning. Compared to the *Classic* approach in Table 5.1, the resulting speed-up is a factor of 2.17. This shows that reusing learned clauses from hard faults helps to improve the overall performance.

Note that *all* possible faults in the circuits where classified by the SAT-based approach. But the overhead of generating a SAT instance only pays off for faults that are hard to classify. In this case, this overhead occurs even for the large number of "easy-to-detect" faults that could be classified much more efficiently by random simulation. Therefore, the overall run time could not be improved in some cases.

Table 5.3: Results for industrial circuits

		Classic		Gate-inp+dynamic	
Circuit	Targets	Ab.	Time	Ab.	Time
p77k	126,338	0	4,487	0	3,832
p80k	181,160	12	24,703	0	12,116
p88k	133,891	2	13,572	0	5,755
p99k	140,633	63	26,343	19	15,397
p177k	260,812	6,974	372,546	236	95,452
p462k	616,735	6,232	309,828	19	62,921
p565k	1,317,213	4,306	495,167	540	284,235
p1330k	1,441,878	132	166,791	14	221,060

Finally, results for industrial benchmark circuits are reported in Table 5.3. The number of faults after collapsing is reported in the second column.

The classical algorithm without learning is compared to the algorithm that combines gate-input-partitioning with dynamic learning. The number of faults that were aborted are reported in column *Ab.* Column *Time* gives the total run time.

The results show that the learning techniques significantly improve the robustness of SAT-based ATPG. A large number of faults was aborted by the classical SAT algorithm. In contrast, only a few aborted faults remain when learning is applied. Moreover, the run time decreases in most cases. The improvement even reaches a factor of 4.9 for circuit p177k. The run time was only increased for p1330k, but at the same time, the number of aborted faults was reduced significantly. This shows that storing learned information is essential to classifying hard faults.

Overall, the performance of SAT-based ATPG can be significantly improved. The combination of gate-input-partitioning and dynamic circuit-based learning especially boosts robustness. The run time is reduced on average and the number of aborted faults is reduced for all benchmarks considered.

5.5 Summary

We have presented an extension to the SAT-based ATPG engine to embed learning strategies. Both paradigms, i.e. incremental SAT and circuit-based learning, have been exploited. For the more difficult case of circuit-based learning, the correctness of the technique has been proven. Experimental results show an improved robustness on large industrial benchmarks.

The next step is the tight integration with classical ATPG engines. In this context, the SAT-based tool can be used to efficiently handle faults that are hard to classify using other techniques. By reusing learned information for the other engines, e.g. FAN, the overall performance can be further improved.

Additionally, the extension of the learning techniques to other fault models, such as the path delay fault model or the bridging fault model, is of interest.

Chapter 6

Multiple-Valued Logic

All circuits considered so far have been Boolean circuits, i.e. all signals and gates can assume one of the Boolean values 0 or 1. However, to use SAT-based ATPG in an industrial environment it is insufficient to consider only Boolean values. Since industrial circuits contain elements that have non-Boolean behavior, ATPG tools for these circuits have to handle these kinds of gates as well.

In this chapter,[1] a brief overview of particular properties of industrial circuits, especially for elements that assume states which cannot be modeled by Boolean values is given. To apply ATPG to those circuits, a four-valued logic is presented in Section 6.1. Furthermore, it is explained in detail how to find an efficient Boolean encoding for this logic. This is done by comparing possible encodings with respect to their sizes in order to find the most suitable one.

Industrial circuits may contain multi-input gates, i.e. gates with more than two inputs. In Section 6.2, the problem of transforming these gates into CNF using four-valued logic is discussed and an approach is given which overcomes the described problems. Experimental results are presented in Section 6.3. In the last section of this chapter, a summary is given.

6.1 Four-Valued Logic

Section 6.1.1 describes the specifics of industrial circuits and introduces values modeling non-Boolean states. Section 6.1.2 presents a multiple-valued logic that can model the additional values and discusses the use of a

[1] A preliminary version of Section 6.1 has been published in [36], whereas parts of Section 6.2 have been published in [100].

Boolean encoding. The efficiency of different Boolean encodings is treated in Section 6.1.3, and a concrete Boolean encoding used throughout this book is presented in Section 6.1.4.

6.1.1 Industrial Circuits

For practical purposes, it is not sufficient to consider only the Boolean values 0 and 1 during test pattern generation as has been done in earlier approaches (e.g. [95]). There are two main reasons.

First, industrial circuits usually contain tri-state elements. Therefore, besides the basic gates shown in Figure 6.1, tri-state elements may also occur in a circuit. These are used if a single signal is driven by multiple sources, e.g. bus structures. Besides the Boolean values 0 and 1, tri-state elements can assume another value, Z, modeling the state of high impedance. These gates behave as follows:

- **BUSDRIVER**, Inputs: a, b, Output: c
  Function: $c = \begin{cases} \text{Z}, & \text{if } a = 0 \\ \text{b}, & \text{if } a = 1 \end{cases}$

- **BUS**, Inputs: $a_1, \ldots, a_n$, Output c
  Function: $c = \begin{cases} \text{Z}, & \text{if } a_1 = \ldots = a_n = Z \\ 0, & \text{if } \exists i \in \{1, \ldots, n\}\ a_i = 0 \\ 1, & \text{if } \exists i \in \{1, \ldots, n\}\ a_i = 1 \end{cases}$
  Note, that the output value is not defined if there are inputs with value 0 and inputs with value 1.

- **BUS0** behaves as BUS, but assumes the value 0 if not being driven.

- **BUS1** behaves as BUS, but assumes the value 1 if not being driven.

Environment constraints that are applied to a circuit are another problem. In industrial practice, the circuit can be embedded in a larger environment. As a result, some inputs of the circuit may not be controllable (see column *Fix* in Table 2.2 in Section 2.5). The value of such a non-controllable

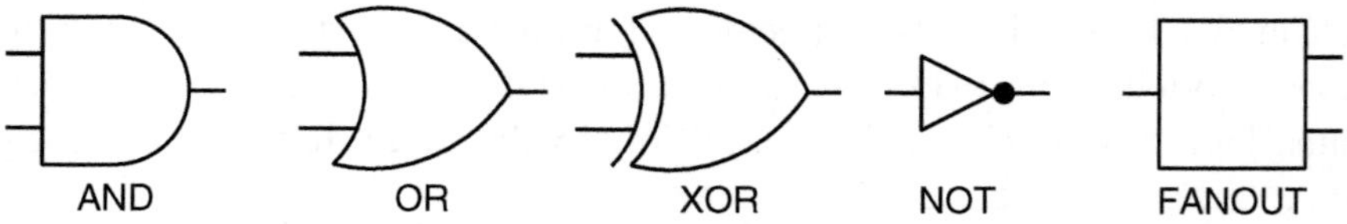

Figure 6.1: Basic gates (replicated from Figure 2.1)

input is assumed to be unknown, denoted by U. Unknown values have to be specially considered during test pattern generation.

Note, that unknown values are not the same as don't care values. Don't care values are allowed to be assigned arbitrarily. Unknown values force a signal to be unassigned during the ATPG process.

For more information about classical algorithms working on industrial circuits using multiple-valued logic, please refer to [103].

6.1.2 Boolean Encoding

From a modeling point of view, a tri-state element could be transformed into a Boolean structure with the same functionality, e.g. by inserting multiplexers. But during test pattern generation, additional constraints apply to signals driven by tri-state elements. For example, no two drivers must drive the signal with opposite values or if all drivers are in the high impedance state, the driven signal has an unknown value (U). The value Z is used to properly model these constraints and the transition function of tri-state elements.

To model unknown values, the logic value U is used. This has to be encoded explicitly in the SAT instance, since otherwise the SAT solver would assign Boolean values to non-controllable inputs.

A four-valued logic $L_4 = \{0, 1, Z, U\}$ can be used to address the above requirements. To apply a Boolean SAT solver to a problem formulated in L_4, the problem has to be transformed into a Boolean problem. Therefore, each signal of the circuit is encoded by two Boolean variables.[2] One encoding out of the $4! = 24$ possible mappings of four values onto two Boolean values has to be chosen. The chosen encoding determines which clauses are needed to model particular gates. This, in turn, influences the size of the resulting SAT instance and the efficiency of the SAT search.

All possible encodings are summarized in Tables 6.1a–c. The two Boolean variables are denoted x and $\overline{x}$, the letters a and b are placeholders for Boolean values. The following gives the interpretation of the tables more formally:

- A signal s is encoded by the two Boolean variables c_s and c_s^*.
- $x \in \{c_s, c_s^*\}, \overline{x} \in \{c_s, c_s^*\} \setminus \{x\}$.
- $a \in \{0, 1\}, \overline{a} \in \{0, 1\} \setminus \{a\}$.
- $b \in \{0, 1\}, \overline{b} \in \{0, 1\} \setminus \{b\}$.

[2]Here, a logarithmic encoding was chosen because it requires the smallest number of Boolean variables to encode a value from the four-valued logic.

Table 6.1: Boolean encodings

(a) Set 1

s	x	$\overline{x}$
0	a	b
1	a	$\overline{b}$
U	$\overline{a}$	b
Z	$\overline{a}$	$\overline{b}$

(b) Set 2

s	x	$\overline{x}$
0	a	b
1	a	$\overline{b}$
U	$\overline{a}$	$\overline{b}$
Z	$\overline{a}$	b

(c) Set 3

s	x	$\overline{x}$
0	a	b
1	$\overline{a}$	$\overline{b}$
U	$\overline{a}$	b
Z	a	$\overline{b}$

(d) Example: Set 1, $a = b = 0,\ x = c_s$

s	c_s	c_s^*
0	0	0
1	0	1
U	1	0
Z	1	1

Example 13 *Consider Set 1 as defined in Table 6.1a and the following assignment:* $a = 0,\ b = 0,\ x = c_s$*. Then, the encoding in Table 6.1d results.*

Thus, a particular encoding is determined by choosing values for a, b and x. Each table defines a set of eight encodings.

Note, that for encodings in Set 1 or Set 2 one Boolean variable is sufficient to decide if the value of s is in the Boolean domain, i.e. in $\{0, 1\}$, or in the non-Boolean domain, i.e. in $\{U, Z\}$. In contrast, encodings in Set 3 do not have this property. This observation will be important when the efficiency of a particular encoding for SAT solving is considered.

Since during ATPG, a difference between the faulty circuit and the fault free circuit has to be found, it is important to know under which circumstances two values are different. If Boolean logic is used, this is straightforward to evaluate. In four-valued logic, however, the classification is more complicated. The following example identifies the problem in detail.

Example 14 *Let* a *and* b *be two variables in four-valued logic. First, assume the assignments*

$$a = 1 \quad \textit{and} \quad b = 1$$

hold. In that case, it is easy to see that both variable values are equal. In the same way, it is straightforward to see, that the assignments

$$a = 1 \quad \textit{and} \quad b = 0$$

result in different values.

Finally, assume the assignments

$$a = 1 \quad \textit{and} \quad b = U$$

Table 6.2: Overview on the differentiation of values in four-valued logic

Equal		Different		Undecidable	
Value 1	Value 2	Value 1	Value 2	Value 1	Value 2
0	0	0	1	0	U
1	1	0	Z	1	U
Z	Z	1	0	U	0
		1	Z	U	1
		Z	0	U	U
		Z	1	U	Z
				Z	U

hold. Here, the value of variable b is unknown, i.e. the actual value of b during the post production test is not predictable. The statement

$$a \text{ and } b \text{ are } \begin{cases} \text{equal} & b = 1 \\ \text{different} & b \in \{0, Z\} \end{cases}$$

holds.

As a result of the unknown values, there is a third class of relationship between two values: Besides *equality* and *difference* the relationship *undecidability* can occur.

A fault can only be observed if a difference between the faulty circuit and the correct circuit can be guaranteed. Therefore, it is insufficient that two variables are not equal – it is necessary that they are explicitly different.

Table 6.2 gives the classification with respect to equality, difference and undecidability for each possible combination of two variables in four-valued logic.

6.1.3 Encoding Efficiency

The clauses to model a specific gate type can be determined if a particular encoding and the truth table of the gate are given. This is done in a manner analogous to the procedure in Section 3.3. The set of clauses can be reduced by two-level logic-optimization. The tool ESPRESSO contained in SIS [89] was used for this purpose. For the small number of clauses for the basic gate types, ESPRESSO is capable of calculating an optimal representation. The following example illustrates the process.

Example 15 *In Table 6.3a, the truth table of an AND gate $s = t \cdot u$ over $\{0, 1, Z, U\}$ is shown. The truth table is mapped onto the Boolean domain*

Table 6.3: AND gate over $\{0, 1, Z, U\}$

(a) Four-valued

t	u	s
0	–	0
–	0	0
1	1	1
U	$\neq 0$	U
Z	$\neq 0$	U
$\neq 0$	U	U
$\neq 0$	Z	U

(b) Encoded

c_t	c_t^*	c_u	c_u^*	c_s	c_s^*
0	0	–	–	0	0
–	–	0	0	0	0
0	1	0	1	0	1
1	0	$\neq 0\ 0$		1	0
1	1	$\neq 0\ 0$		1	0
$\neq 0\ 0$		1	0	1	0
$\neq 0\ 0$		1	1	1	0

Table 6.4: Number of clauses for each encoding

Set	NAND	NOR	AND	BUS	BUS0	BUS1	BUSD.	XOR	NOT	OR	All
1	8	9	9	10	11	10	9	5	5	8	100
2	9	8	8	10	10	11	9	5	5	9	100
3	11	11	11	8	9	9	11	5	6	11	108

using the encoding from Example 13. The encoded truth table is shown in Table 6.3b (for compactness the notation "$\neq 0\ 0$" is used to denote that at least one of two variables must be different from 0; "–" denotes "don't care"). A CNF is extracted from this truth table and optimized by ESPRESSO.

Results for all possible encodings are presented in Table 6.4. For each gate type, the number of clauses needed to model the gate's function are given. Besides the well-known Boolean gates (*AND*, *OR*, ...), the non-Boolean gates *BUSDRIVER*, *BUS0* and *BUS1* described in Section 6.1.1 are also considered. The last column *All* in the table gives the sum of the numbers of clauses for all gate types.

All encodings of a given set lead to clauses that are isomorphic to each other. By mapping the polarity of literals and the choice of variables, the other encodings of the set are retrieved. In particular, Boolean gates are modeled efficiently by encodings from Set 1 and Set 2. The sum of clauses needed for all gates is equal for both sets. One difference for example is that the encodings of one set are more efficient for *NAND* gates, while the encodings of the other set are more efficient for *NOR* gates.

Both gate types occur with similar frequency in the industrial circuits as shown in Table 6.5. The same observation is true for the other gates where the efficiency of the encodings differs. Therefore, no significant trade-off for the encodings occurs on the benchmarks.

Table 6.5: Number of gates for each type

Circ.	IN	OUT	FANO.	NOT	AND	NAND	OR	NOR	BUS	BUSD.
p44k	2,914	2,231	6,763	16,869	12,365	528	5,484	1,128	0	0
p88k	4,712	4,564	14,560	20,913	27,643	2,838	16,941	5,883	144	268
p177k	11,275	10,508	25,394	48,582	49,911	5,707	30,933	5,962	0	560

In contrast, more clauses are needed to model Boolean gates if an encoding of Set 3 is used. At the same time, this encoding is more efficient for non-Boolean gates. In most circuits, the number of non-Boolean gates is usually much smaller than the number of Boolean gates. Therefore, more compact SAT instances will result if an encoding from Set 1 or Set 2 is used.

The behavior of the SAT solver does not necessarily depend on the size of the SAT instance, but if the same problem is encoded in a much smaller instance, better performance of the SAT solver can be expected. These hypotheses are strengthened by the experimental results reported in Section 6.3.

To close this overview on the efficiency of Boolean encodings for the four-valued logic, a concrete encoding used to apply ATPG on industrial circuits is given below. This Boolean encoding will be used throughout this book for stuck-at fault test pattern generation.

6.1.4 Concrete Encoding

Table 6.6 shows the Boolean encoding of the four-valued logic L_4 used in this work. It is an encoding from Set 2 (see Table 6.1b) where c_s^* indicates whether the value is in the Boolean domain or not, i.e. the value 0 means the Boolean domain, whereas 1 indicates one of the additional values U or Z. During experiments, this encoding worked well for industrial circuits.

Example 16 *In Table 6.7, the complete CNF for an AND gate with inputs a and b and output o is depicted. In total, 15 variables are needed, namely:*

- a_c, b_c, o_c – *first variable for the correct circuit*
- a_c^*, b_c^*, o_c^* – *second variable for the correct circuit*
- a_f, b_f, o_f – *first variable for the faulty circuit*
- a_f^*, b_f^*, o_f^* – *second variable for the faulty circuit*
- a_d, b_d, o_d – *D-chain variable*

Table 6.6: Boolean encoding of L_4 used in the following

s	c_s	c_s^*
0	0	0
1	1	0
U	1	1
Z	0	1

Table 6.7: CNF for an AND gate using L_4

$$
\begin{array}{llll}
1 & (\overline{a}_c + \overline{b}_c + o_c) & \cdot\, (a_c^* + b_c^* + \overline{o}_c^*) & \cdot\, (o_c + \overline{o}_c^*)\, \cdot \\
 & (a_c + a_c^* + \overline{o}_c) & \cdot\, (b_c + b_c^* + \overline{o}_c) & \cdot\, (\overline{a}_c^* + \overline{b}_c + o_c^*)\, \cdot \\
 & (\overline{a}_c + \overline{b}_c^* + o_c^*) & \cdot\, (\overline{a}_c^* + \overline{b}_c^* + o_c^*)\, \cdot & \\
2 & (\overline{a}_f + \overline{b}_f + o_f) & \cdot\, (a_f^* + b_f^* + \overline{o}_f^*) & \cdot\, (o_f + \overline{o}_f^*)\, \cdot \\
 & (a_f + a_f^* + \overline{o}_f) & \cdot\, (b_f + b_f^* + \overline{o}_f) & \cdot\, (\overline{a}_f^* + \overline{b}_f + o_f^*)\, \cdot \\
 & (\overline{a}_f + \overline{b}_f^* + o_f^*) & \cdot\, (\overline{a}_f^* + \overline{b}_f^* + o_f^*)\, \cdot & \\
3 & (\overline{o}_d + \overline{o}_c + \overline{o}_f) & \cdot\, (\overline{o}_d + \overline{o}_c^* + \overline{o}_f^*) & \cdot\, (\overline{o}_d + \overline{o}_f + \overline{o}_f^*)\, \cdot \\
 & (\overline{o}_d + \overline{o}_c + \overline{o}_c^*) & \cdot\, (\overline{o}_d + \overline{o}_c + \overline{o}_c^* + \overline{o}_f + \overline{o}_f^*)\, \cdot & \\
4 & (\overline{a}_d + o_d) & \cdot\, (\overline{b}_d + o_d) &
\end{array}
$$

As can be seen, the CNF in Table 6.7 is divided into four parts. The first and the second part represent the CNF for the AND gate in the correct circuit and in the faulty circuit, respectively. The CNF in the third part makes sure that the values of both circuits differ if the gate is on a D-chain, i.e. those clauses encode the property

$$o_d \rightarrow \hat{o}_c \neq \hat{o}_f,$$

where $\hat{o}_c$ *and* $\hat{o}_f$ *denote the correct and the faulty variable in four-valued logic, respectively. Finally, the clauses depicted in the last part describe the property*

$$a_d \rightarrow o_d \quad \text{and} \quad b_d \rightarrow o_d.$$

This part of the CNF ensures that the gate itself is on a D-chain if one of its predecessors is on a D-chain.

It is easy to see that only the first two paragraphs have to be modified when the CNF should describe another gate type. The D-chain clauses do not change since they are independent of the gate type.

Table 6.8: CNF for a BUSDRIVER using L_4

$(b^*_c + \overline{a}_c + a^*_c + \overline{c}^*_c)$	$\cdot\ (b_c + a^*_c + \overline{c}_c)$	$\cdot\ (a_c + a^*_c + \overline{c}_c)\ \cdot$
$(\overline{a}^*_c + c_c)$	$\cdot\ (\overline{b}_c + \overline{a}_c + c_c)$	$\cdot\ (\overline{a}^*_c + \overline{c}^*_c)\ \cdot$
$(a_c + c^*_c)$	$\cdot\ (\overline{b}^*_c + c^*_c)$	

Table 6.9: CNF for a BUS using L_4

$(a_c + a^*_c + \overline{c}_c + c^*_c)$	$\cdot\ (b^*_c + c_c + \overline{c}^*_c)$	$\cdot\ (a_c + b_c + \overline{c}_c)\ \cdot$
$(b_c + b^*_c + \overline{c}_c + c^*_c)$	$\cdot\ (a^*_c + b_c + \overline{b}^*_c + \overline{c}^*_c)$	$\cdot\ (a_c + \overline{a}^*_c + b^*_c + \overline{c}^*_c)\ \cdot$
$(\overline{a}_c + a^*_c + \overline{b}_c + b^*_c + \overline{c}^*_c)$	$\cdot\ (\overline{b}_c + \overline{b}^*_c + c^*_c)$	$\cdot\ (\overline{a}_c + \overline{a}^*_c + c^*_c)\ \cdot$
$(\overline{b}_c + c_c)$	$\cdot\ (\overline{a}_c + c_c)$	$\cdot\ (\overline{a}^*_c + \overline{b}^*_c + c^*_c)$

Example 17 *Table 6.8 shows the CNF for a BUSDRIVER where a is the select input, b is the data input and c is the output signal. Table 6.9 presents the CNF for a BUS element with inputs a, b and output c. Note, that here only the clauses for the good circuit are depicted. Generating clauses for the faulty circuit is straightforward using the corresponding variables. Clauses to model the D-chain are created in the manner explained above.*

6.2 Multi-input Gates

Until now, when transforming a single gate into CNF, it has been assumed implicitly that each gate has exactly two inputs (except buffer and inverter gates). Industrial circuits, however, contain gates with more than two inputs. These gates are called in the following *multi-input gates*. In former approaches, during CNF transformation, multi-input gates have been decomposed into a sequence of two-input gates [91, 95].

Table 6.10 shows the distribution of multi-input gates in some industrial circuits from NXP Semiconductors. In each column, the accumulated number of AND, NAND, OR and NOR gates with the respective number of inputs is shown.

6.2.1 Modeling of Multi-input Gates

In this section, different ways to model a multi-input gate are studied. For the sake of convenience, in the following, an n-input gate is called an *n-gate*.

In ATPG tools, multi-input gates are often modeled as cascades of 2-gates (see e.g. [91]). The formal definition of this construction is given in the

Table 6.10: Distribution of n-input gates

Circuit	2	3	4	5	6	7	8
p44k	14,461	3,257	1,702	0	0	0	85
p80k	43,487	9,916	5,167	0	0	0	0
p88k	48,594	4,162	549	0	0	0	0
p99k	47,608	5,191	338	2	0	2	15
p177k	84,792	6,324	1,397	0	0	0	0
p462k	203,050	14,050	2,726	461	0	0	0
p565k	376,450	18,531	1,982	0	0	0	0
p1330k	421,658	44,014	4,076	0	0	0	0

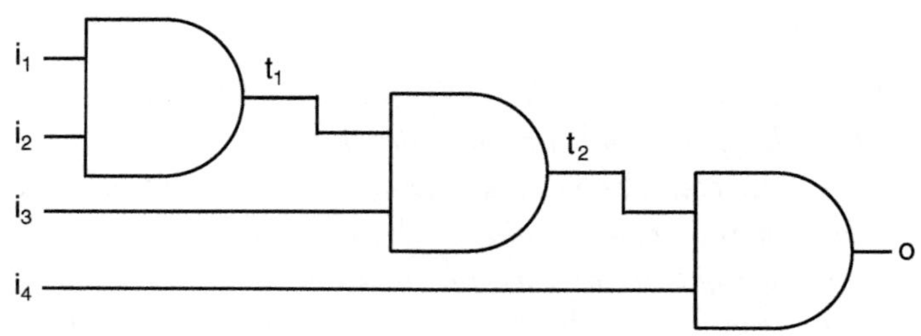

Figure 6.2: 4-AND gate modeled by a sequence of three 2-AND gates

following: Let $\odot$ be the gate's function with inputs $i_1, \ldots, i_n$ (where $n > 2$) and output o. Then o is calculated as follows:

$$\begin{aligned} t_0 &:= i_1 \\ t_j &:= i_{j+1} \odot t_{j-1} \quad \text{for } j = 1, \ldots, n-1 \\ o &:= t_{n-1} \end{aligned}$$

where $t_1, \ldots, t_{n-2}$ are connections between the 2-gates. This approach is illustrated for a 4-AND gate in Figure 6.2.

Due to the auxiliary connections (in Figure 6.2 denoted by t_2 and t_3) there is an overhead of $n - 1$ variables in Boolean logic and $2 \cdot (n - 1)$ variables in the four-valued logic.

These auxiliary variables can be avoided if an n-gate is modeled as one single gate. However, the number of clauses needed to model an n-gate in four-valued logic grows exponentially. Table 6.11 shows the CNF sizes for n-input AND and OR gates. The columns *2-input*, *Multi-input* and *Bounded* present the numbers for the approach "divide the n-gate into $(n-1)$ 2-gates", "use normal n-gate" and "use bounded multi-input gates", respectively. The bounded multi-input approach is explained in detail in the next

Table 6.11: CNF sizes for n-gates occurring in industrial circuits

	2-input		Multi-input		Bounded	
Gate	Variables	Clauses	Variables	Clauses	Variables	Clauses
2-AND	6	8	6	8	6	8
3-AND	10	16	8	13	8	13
4-AND	14	24	10	22	10	22
5-AND	18	32	12	39	12	39
6-AND	22	40	14	72	16	47
7-AND	26	48	16	137	18	52
8-AND	30	56	18	266	20	61
2-OR	6	9	6	9	6	9
3-OR	10	18	8	15	8	15
4-OR	14	27	10	25	10	25
5-OR	18	36	12	43	12	43
6-OR	22	45	14	77	16	52
7-OR	26	54	16	143	18	58
8-OR	30	63	18	273	20	68

section. Columns entitled *Variables* and *Clauses* give the number of variables and clauses, respectively. Note, that the sizes for AND and OR gates are equal to the sizes for NOR and NAND gates, respectively.

As basic gates, the CNF sizes of the 2-AND and the 2-OR gates are equal in all three cases. At each input level, the number of variables in the 2-input approach grows by four, whereas, in the multi-input approach, only two additional variables in each level are needed. So the difference of the two approaches is 12 variables for an 8-gate.

However, the number of clauses needed to model an AND gate or an OR gate with more than five inputs exceeds the number needed in the 2-input approach. More than four times the clauses for an 8-OR gate or an 8-AND gate are required. The reason is the exponential growth in the number of clauses needed. To model an n-AND $2^n + n + 2$ clauses are needed and an n-OR requires $2^n + 2 \cdot n + 1$ clauses.

An example of these two approaches is shown in Figure 6.3. The 4-AND gate from Figure 6.2 is converted into a CNF, where in Figure 6.3a the cascaded 2-AND approach is used and in Figure 6.3b the 4-AND gate was modeled as a single gate. In the multi-input approach, the CNF is smaller: Two clauses and four variables less than in the 2-input approach are needed. This corresponds to line three in Table 6.11.

$$
\begin{array}{lll}
(\bar{i}_1 + \bar{i}_2 + t_1) & \cdot (i_1^* + i_2^* + \bar{t}_1^*) & \cdot (t_1 + \bar{t}_1^*) \cdot \\
(i_1 + i_1^* + \bar{t}_1) & \cdot (i_2 + i_2^* + \bar{t}_1) & \cdot (\bar{i}_1^* + \bar{i}_2 + t_1^*) \cdot \\
(\bar{i}_1 + \bar{i}_2^* + t_1^*) & \cdot (\bar{i}_1^* + \bar{i}_2^* + t_1^*) \cdot & \\
(\bar{t}_1 + \bar{i}_3 + t_2) & \cdot (t_1^* + i_3^* + \bar{t}_2^*) & \cdot (t_2 + \bar{t}_2^*) \cdot \\
(t_1 + t_1^* + \bar{t}_2) & \cdot (i_3 + i_3^* + \bar{t}_2) & \cdot (\bar{t}_1^* + \bar{i}_3 + t_2^*) \cdot \\
(\bar{t}_1 + \bar{i}_3^* + t_2^*) & \cdot (\bar{t}_1^* + \bar{i}_3^* + t_2^*) \cdot & \\
(\bar{t}_2 + \bar{i}_4 + o) & \cdot (t_2^* + i_4^* + \bar{o}^*) & \cdot (o + \bar{o}^*) \cdot \\
(t_2 + t_2^* + \bar{o}) & \cdot (i_4 + i_4^* + \bar{o}) & \cdot (\bar{t}_2^* + \bar{i}_4 + o^*) \cdot \\
(\bar{t}_2 + \bar{i}_4^* + o^*) & \cdot (\bar{t}_2^* + \bar{i}_4^* + o^*) &
\end{array}
$$

(a) 4-AND consisting of 2-ANDs

$$
\begin{array}{ll}
(\bar{i}_1 + \bar{i}_2 + \bar{i}_3 + \bar{i}_4^* + o^*) & \cdot (\bar{i}_1 + \bar{i}_2 + \bar{i}_3^* + \bar{i}_4 + o^*) \cdot \\
(\bar{i}_1 + \bar{i}_2^* + \bar{i}_3 + \bar{i}_4 + o^*) & \cdot (\bar{i}_1^* + \bar{i}_2 + \bar{i}_3 + \bar{i}_4 + o^*) \cdot \\
(\bar{i}_1 + \bar{i}_2 + \bar{i}_3^* + \bar{i}_4^* + o^*) & \cdot (\bar{i}_1 + \bar{i}_2^* + \bar{i}_3 + \bar{i}_4^* + o^*) \cdot \\
(\bar{i}_1^* + \bar{i}_2 + \bar{i}_3 + \bar{i}_4^* + o^*) & \cdot (\bar{i}_1 + \bar{i}_2^* + \bar{i}_3^* + \bar{i}_4 + o^*) \cdot \\
(\bar{i}_1^* + \bar{i}_2 + \bar{i}_3^* + \bar{i}_4 + o^*) & \cdot (\bar{i}_1^* + \bar{i}_2^* + \bar{i}_3 + \bar{i}_4 + o^*) \cdot \\
(\bar{i}_1 + \bar{i}_2^* + \bar{i}_3^* + \bar{i}_4^* + o^*) & \cdot (\bar{i}_1^* + \bar{i}_2 + \bar{i}_3^* + \bar{i}_4^* + o^*) \cdot \\
(\bar{i}_1^* + \bar{i}_2^* + \bar{i}_3 + \bar{i}_4^* + o^*) & \cdot (\bar{i}_1^* + \bar{i}_2^* + \bar{i}_3^* + \bar{i}_4 + o^*) \cdot \\
(\bar{i}_1^* + \bar{i}_2^* + \bar{i}_3^* + \bar{i}_4^* + o^*) & \cdot (i_4 + i_4^* + \bar{o}) \cdot \\
(i_3 + i_3^* + \bar{o}) & \cdot (i_2 + i_2^* + \bar{o}) \cdot \\
(i_1 + i_1^* + \bar{o}) & \cdot (o + \bar{o}^*) \cdot \\
(\bar{i}_1 + \bar{i}_2 + \bar{i}_3 + \bar{i}_4 + o) & \cdot (i_1^* + i_2^* + i_3^* + i_4^* + \bar{o}^*)
\end{array}
$$

(b) 4-AND as one single gate

Figure 6.3: Clauses for the four-valued 4-AND gate

6.2.2 Bounded Multi-input Gates

The advantages and drawbacks of modeling multi-input gates for many inputs were described above. In the 2-input approach, the number of variables grows significantly. In the multi-input approach, the number of variables grows slightly, but there is an exponential growth of clauses which is smaller in the 2-input approach. Up to five inputs the number of clauses is acceptable. Therefore a *bounded multi-input* approach is proposed. This is a combination where the multi-input approach is used, but the number of inputs per gate is limited.

According to Table 6.11, the input number is set to five. To model a gate with more than five inputs, too many clauses are required while, on the other hand, the variable saving in gates with less than five inputs are low. In the bounded multi-input approach, gates with more than five inputs are divided into sequences of 5-gates.

6.2.3 Clause Generation

Figure 6.4 shows the general flow of the clause generation procedure. Consider the dashed box on the left side: First, the truth table of a gate's function is created. This table is translated into a CNF using a dedicated script *table2cnf*. This CNF is minimized by ESPRESSO [89]. Since the Boolean function is small enough, an optimal minimization algorithm can be applied. The script *pla2cnf* converts the minimized CNF into C++ code that adds clauses to the SAT solver. This function is independent of the SAT solver used since an abstract interface is used.

This work flow is applied once for each gate type and for each number of gate inputs. Each pass creates a function in C++ code which is exported into a library. This library is included in the ATPG tool. For each gate occurring in the circuit that has to be added to the SAT solver, the respective function in the library is called.

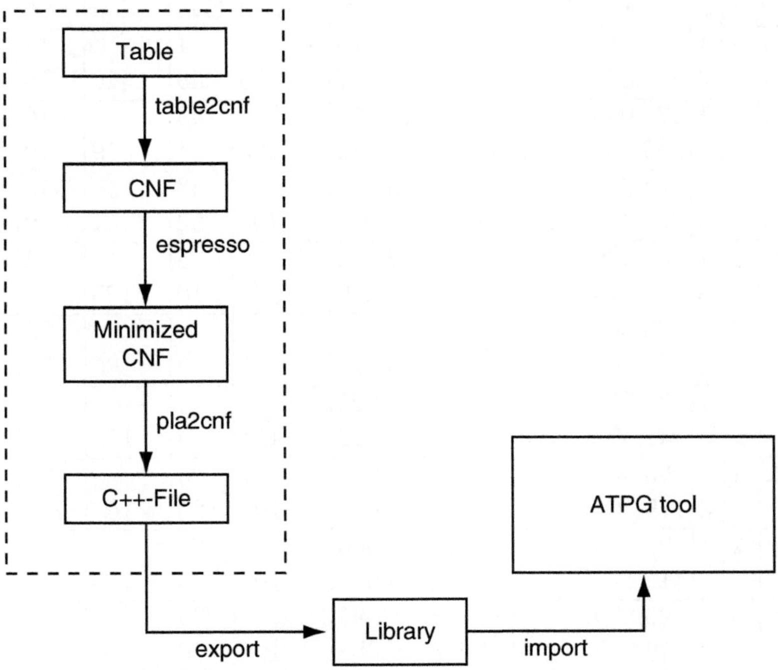

Figure 6.4: Clause generation working flow

6.3 Experimental Results

Experimental results are given in this section. First, the influence of the chosen Boolean encoding of L_4 is shown by example. Afterwards, in Section 6.3.2, the experimental results of the application to multi-input gates is presented. Statistical information about the industrial circuits can be found in Section 2.5.

6.3.1 Four-Valued Logic

The experiments presented in this subsection were carried out on an AMD Athlon XP 3000+ (2.1 GHz, 1,024 MByte RAM, GNU/Linux). zChaff [82], 2004 version, was used as the SAT solver.

The results in Table 6.12 help to evaluate the influence of different encodings. As explained in Section 6.1.2, there are 24 possibilities to choose a particular encoding. As discussed in Section 6.1.3 a significant trade-off in run time or memory needs cannot be expected for the encodings of Set 1 and

Table 6.12: Memory and run time for different encodings

Circ	#	E	Cls	%Cls	Vars	Mem	%Mem	Eqn	%Eqn	SAT	%SAT
p44k	1	1	173,987		56,520	13,713		41		14	
		3	220,375	127	56,520	14,087	103	49	120	78	557
p44k	2	1	174,083		56,542	13,713		43		16	
		3	220,493	127	56,542	14,088	103	51	119	79	494
p44k	3	1	174,083		56,542	13,713		43		15	
		3	220,493	127	56,542	14,088	103	52	121	79	527
p88k	1	1	33,406		10,307	2,824		8		4	
		3	41,079	123	10,307	3,410	121	10	125	7	175
p88k	2	1	33,501		10,328	2,824		9		4	
		3	41,188	123	10,328	3,411	121	9	100	8	200
p88k	3	1	33,517		10,289	2,825		8		8	
		3	41,321	123	10,289	3,412	121	9	113	8	100
p177k	1	1	96,550		34,428	8,900		23		23	
		3	119,162	123	34,428	9,082	102	25	107	247	1074
p177k	2	1	96,536		34,425	8,900		25		28	
		3	119,145	123	34,425	9,082	102	29	116	234	836
p177k	3	1	96,550		34,428	8,899		25		20	
		3	119,162	123	34,428	9,082	102	29	116	237	1185

Set 2. Therefore, we chose one encoding from Set 1 and one encoding from the remaining Set 3. The industrial benchmarks already shown in Table 6.5 were considered.

Table 6.12 presents results for three faults on each circuit. The name of the circuit (column *Circ*), an id number for the fault (column #) and the type of encoding used, i.e the Set (column E) are reported. The memory needs were measured by the number of clauses (column *Cls*), the number of variables (column *Vars*) and memory consumption in kByte (column *Mem*). The run times are measured for the generation of the CNF formula (column *Eqn*) and solving the formula (column *SAT*). In all cases, the overhead of the encoding from Set 3 over the encoding from Set 1 is shown as a percentage.

For all the test cases, the encoding from Set 1 performs significantly better than the encoding from Set 3. As can be expected from the number of clauses needed per gate as shown in Table 6.4, the memory needs are larger for the encoding from Set 3. The number of variables does not depend on the encoding, but on the number of gates in the circuit and remains the same. The influence of the encoding on the run time is even more remarkable than the influence on the memory needs. The run time for the third fault in p177k is almost 12 times faster if the encoding from Set 1 is applied.

6.3.2 Multi-input Gates

The experiments for the techniques presented in Section 6.2 were carried out with an improved version of PASSAT. All experiments were carried out on an Intel Xeon system (3 GHz, 32,768 MByte RAM, GNU/Linux). As benchmarks, industrial circuits from NXP Semiconductors were used. MiniSat [29] v1.14 was used as the SAT solver.

The results are presented in Table 6.13. Column *Circuit* shows the circuit's name. The columns *2-input* and *Bounded multi-input* show the results of the two approaches for clause encoding, where the columns *Aborted* and *Run time* give the number of timeouts during the search and the total run time of the entire ATPG search, respectively. An abort occurs when the CNF for a fault is not solvable within 20 CPU seconds.

As can be seen, on all circuits, the results of the bounded multi-input approach are better than the results of the 2-input approach. On some circuits, the new approach even yields substantially improved results (e.g. p44k) and on some circuits the gain is small (e.g. p88k).

This is explained by Table 6.10. Consider the relative number of multi-input gates (with respect to the number of all gates). In circuits with considerable speed-up, this number is high. In these cases, fewer variables are

Table 6.13: Experimental results

Circuit	2-input		Bounded multi-input	
	Aborted	Run time	Aborted	Run time
p44k	2,583	25:11 h	0	2:18 h
p80k	1	52:24 min	1	42:58 min
p88k	0	12:13 min	0	11:41 min
p99k	0	9:07 min	0	8:41 min
p177k	1,337	13:26 h	941	10:28 h
p462k	155	3:53 h	129	3:31 h
p565k	0	2:42 h	0	2:23 h
p1330k	1	5:28 h	1	4:58 h

Table 6.14: Instance sizes

Circuit	2-input		Bounded multi-input	
	Variables	Clauses	Variables	Clauses
p44k	71,933	221,436	60,446	209,001
p80k	8,308	23,328	7,483	22,697
p88k	5,276	15,951	5,047	15,676
p99k	5,689	16,687	5,301	16,139
p177k	73,506	227,871	69,908	222,516
p462k	7,473	22,346	7,336	22,399
p565k	4,829	15,827	4,663	15,638
p1330k	21,459	64,170	20,580	62,929

needed and therefore, the SAT instance can be solved more easily. In contrast, circuits with small speed-up have a low number of multi-input gates.

Table 6.14 provides further insight. There, an overview of the average size of the SAT instances with respect to the number of variables (column *Variables*) and the number of clauses (column *Clauses*) is given.

It can be seen that the bounded multi-input approach generates SAT instances with fewer variables and (except for p462k) with fewer clauses than the 2-input approach. Besides reducing the run time, this also implies savings in memory requirements.

6.4 Summary

In this chapter, the four-valued logic L_4 applicable to SAT-based ATPG on industrial circuits has been introduced. First, a brief introduction into tri-state elements and unknown values has been given. These are important components of industrial circuits. In order to use Boolean SAT solvers, different Boolean encodings of L_4 have been proposed and a comparison with respect to the resulting size of the SAT instances has been made.

After that, an overview on multi-input gates has been given. Their special treatment has been motivated and an approach of a very compact CNF representation which results in significant run time savings has been presented.

Chapter 7

Improved Circuit-to-CNF Conversion

So far, it has been shown how to encode an ATPG problem into a Boolean satisfiability problem and how to convert a circuit into a SAT instance represented in CNF. The use of multiple-valued logic for handling industrial circuits has also been addressed. In this chapter, improvements to the circuit-to-CNF conversion are proposed so that the approach can cope with large circuits very efficiently.

In the first section of this chapter,[1] the use of a *hybrid logic* is proposed. As mentioned above, the use of the four-valued logic L_4 to handle industrial circuits containing tri-state elements and unknown values creates some overhead – especially in circuits that could completely be modeled with the Boolean logic L_2.

The most straightforward way would be to use L_2 for Boolean circuits and L_4 for non-Boolean circuits, i.e. industrial circuits containing non-Boolean elements. However, this approach is not optimal, because in most industrial circuits only a few gates are non-Boolean, and thus, only a small portion of the entire circuit has to be modeled by the four-valued logic. Using this observation, the use of a hybrid logic is proposed that makes it possible to encode the circuit parts that are influenced by non-Boolean gates with the logic L_4 while the other circuit parts can be modeled in Boolean logic. This results in more compact CNFs and the SAT solving process is then accelerated.

Further, an *incremental instance generation* scheme is proposed in Section 7.2. A detailed analysis of SAT-based ATPG is given by com-

[1]Parts of Sections 7.1 and 7.2 have been published in [26] and [99], respectively.

R. Drechsler et al., *Test Pattern Generation using Boolean Proof Engines*,

paring the time needed to generate a SAT instance with the time needed for solving a SAT instance. It is shown that the build time is not only a significant part of the overall time, but often dominates it. Furthermore, it is shown that testable faults are usually harder to classify than untestable faults.

Based on these observations, an instance generation scheme is proposed where initially only a small portion of the influenced subcircuit is transformed into a CNF. This CNF is augmented as necessary.

Experimental results for the proposed techniques are presented in Section 7.3 and a summary of this chapter is given in Section 7.4.

7.1 Hybrid Logic

This section presents a fast preprocessing step that achieves a reduced size for the CNF by partially modeling the circuit in Boolean logic instead of in L_4. As described in the last chapter, circuits including non-Boolean elements cannot be modeled directly with Boolean logic. Therefore, a Boolean encoding is needed for applying SAT-based algorithms. By applying the encoding, the size of the SAT instance increases significantly.

Table 7.1 shows the number of clauses (columns *Cls*) and literals (columns *Lit*) needed to represent a 2-input AND gate in Boolean logic (column *Boolean*) and in L_4 (column *Four-valued*). Column *∅len* gives the average clause length. Transforming circuits with multiple-valued logic results in larger and often also more difficult to solve SAT instances than transforming circuits containing only Boolean logic.

In most industrial circuits, the number of tri-state elements is very small compared to the number of Boolean gates. The state of high impedance represented as the non-Boolean value Z can only be assumed in those elements.

Table 7.1: CNF size for a 2-input AND gate

	CNF description					
	Boolean			Four-valued		
Constraint	Cls	Lit	∅len	Cls	Lit	∅len
$C_g \equiv (A_g \cdot B_g)$	3	7	2.3	8	23	2.9
$C_f \equiv (A_f \cdot B_f)$	3	7	2.3	8	23	2.9
$C_d \rightarrow (C_g \neq C_f)$	2	6	3.0	5	16	3.2
$C_d \rightarrow (D_d + E_d)$	1	3	3.0	1	3	3.0
Overhead	1.0	1.0	1.0	2.4	2.8	1.1

In the case of propagating the Z-value to a Boolean gate, the value Z is interpreted as an unknown state represented as the non-Boolean value U. Additionally, the unknown state can be assumed by inputs when they are fixed to a non-Boolean value.

For that reason, only those elements that can assume non-Boolean values, should be modeled in L_4. An element of the circuit can assume a non-Boolean value if, and only if,

- The element can handle Z-values, i.e. a tri-state element,
- The element is a (pseudo) primary input of the circuit and is fixed to a non-Boolean value, or
- The element is contained in the output cone of one or more of the above mentioned elements.

All other elements can only assume Boolean values and can be modeled in Boolean logic. Additionally, according to their function, tri-state elements are usually located near the circuit outputs. This results in small output cones for the tri-state elements. Furthermore, the percentage of inputs with unknown state is typically very small. Therefore, only a small subset S of elements has to be modeled in L_4.

Determining the subset S is done in a preprocessing step by analyzing the structure of the circuit and classifying the circuit's elements. Algorithm 2 shows pseudo code for a procedure to determine the elements of the subset S of gates which can assume the non-Boolean values U or Z. A description of the algorithm follows.

For structural classification, a modified depth-first search is applied to the circuit. First, all non-Boolean elements of the circuit (i.e. tri-state elements and inputs fixed to U) are identified and stored in a list (line 3). Each element of the list is successively added to the set S (line 6). Every gate in the fanout cones of those elements must also be an element of S, because a non-Boolean value can be propagated via this gate. For that reason, the successors of gates in S are added to the list (line 9), from which the current gate is deleted (line 17). When the list becomes empty, all gates of the fanout cones of non-Boolean elements are contained in S and the subset S is fully determined.

Additionally, all direct predecessors p of a gate g in S with $p \notin S$ are marked as *transition* (line 14). At those gates, a transition between the different logics occurs, i.e. the output and at least one input of a gate are modeled in different logics. Those transitions must be specially handled to

Algorithm 2 Pseudo code of the structural classification

```
list<gate> l;
set<gate> S;
l.add(all_non_Boolean_elements());
while !l.empty() do
   gate elem = l.first_element();
   S.add(elem);
   for all succ ∈ elem.all_successors() do
      if succ ∉ S & succ ∉ l then
         l.append(succ);
      end if
   end for
   for all pred ∈ elem.all_predecessors() do
      if pred ∉ S & pred ∉ l then
         mark_as_transition(pred);
      end if
   end for
   l.remove(elem);
end while
```

guarantee consistency. To avoid inconsistencies due to the different encoding, each gate marked as a *transition* gets a second variable, which is fixed to 0. Due to reconvergent paths, there is the possibility that gates which are contained in S are also marked as transition. These are ignored when fixing the second variable to 0.

Handling the logic transitions between Boolean logic and L_4 can be done in the straightforward way just mentioned thanks to the chosen Boolean encoding of L_4 (see Section 6.1.4). One of the two Boolean variables determines whether a value is Boolean or not. By fixing this variable to zero, the respective signal can only assume Boolean variables although it is encoded in four-valued logic.

Due to marking the predecessors, the complexity for the structural classification would be quadratic in the number of gates in the worst-case. But in practice, gates only have k predecessors with $k \ll n$, where n is the number of gates. For that reason, the complexity is given by $\mathcal{O}(k \cdot n)$. The structural classification is required only once – prior to the ATPG process – and the extracted information can be used for each fault.

The following example demonstrates how this procedure works.

Example 18 *Consider the part of a circuit shown in Figure 7.1. The gates k and n are tri-state elements and are therefore added to the set S. All successors of these gates can assume non-Boolean values. Therefore, p and q are also added to S.*

Additionally, the gates h, i, l, m and o are marked as transition, because they are not in S but have a successor $s \in S$. In Figure 7.1, the outgoing lines of those gates, which are elements of S are marked bold and have an index 4, whereas outgoing lines of gates which are marked as transition have an index t.

Two Boolean variables are assigned to the following gates:

$$h, i, k, l, m, n, o, p, q$$

Note, that the second Boolean variable of h, i, l, m, o is fixed to 0 because they can only assume Boolean values. For gates (including inputs)

$$a, b, c, d, e, f, g, j$$

only one Boolean variable is needed.

Once all gates are classified, the additional information can be used while generating the CNF. A gate $g \in S$ is modeled in L_4, whereas for each gate $h \notin S$, the Boolean logic L_2 is used. More formally, the CNF Φ_g for each gate g in the circuit can be determined by the following equation:

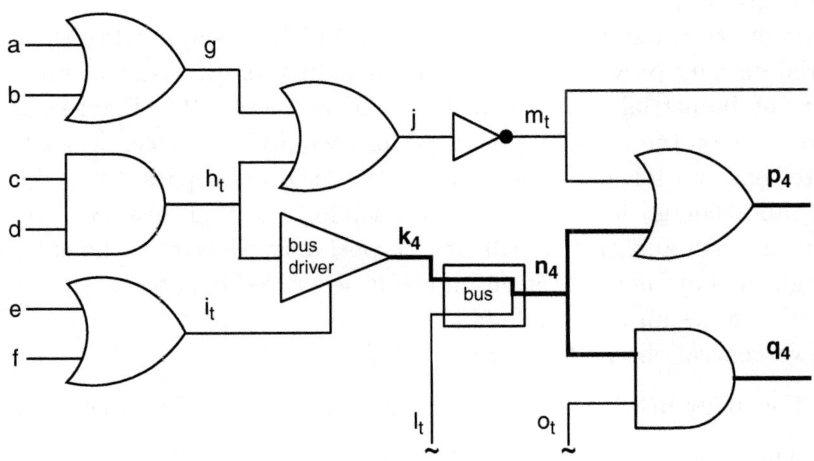

Figure 7.1: Structural classification

$$\Phi_g = \begin{cases} \Phi_g^4, & \text{if } g \in S \\ \Phi_g^2, & \text{if } g \notin S \end{cases}$$

where Φ_g^4 and Φ_g^2 denote the CNF of gate g modeled in L_4 and L_2, respectively.

By using L_2 instead of L_4 where possible, the size of the CNF is decreased. The larger the portion of gates, which only assume Boolean values, the larger is the reduction in size. In addition, there is negligible overhead in run time, because the preprocessing step has to be executed only once before test generation.

7.2 Incremental Instance Generation

In this section, an incremental solving scheme for SAT-based ATPG is proposed. Based on the motivation for this work – which will be given in the form of a detailed run time analysis of PASSAT – a technique to generate only a partial CNF is introduced. If this SAT instance is satisfiable, a test pattern can be derived. Otherwise, the SAT instance is enlarged and a new CNF is generated. Information from previous SAT computations can be reused for the expanded instance.

7.2.1 Run Time Analysis

As mentioned in Chapter 4, SAT-based ATPG consists of two steps: building a SAT instance and solving it. In the following, a detailed analysis of both steps is given.

For the run time analysis, SAT-based ATPG was applied to several industrial circuits provided by NXP Semiconductors. Statistical information about the industrial circuits can be found in Section 2.5. Figure 7.2 gives an overview on the run times required for each SAT instance, i.e. each entry denotes one run for a specific fault. In the diagram, separate run times (in CPU milliseconds) for generating and solving the SAT instance are given on the abscissa and on the ordinate, respectively. Moreover, the entries are distinguished by their classification result, where '+' denotes a testable fault and '×' denotes an untestable fault.

Two general observations can be made:

1. For many instances, the generation time exceeds the solving time.
2. The solving time of testable instances exceeds the solving time of untestable instances significantly.

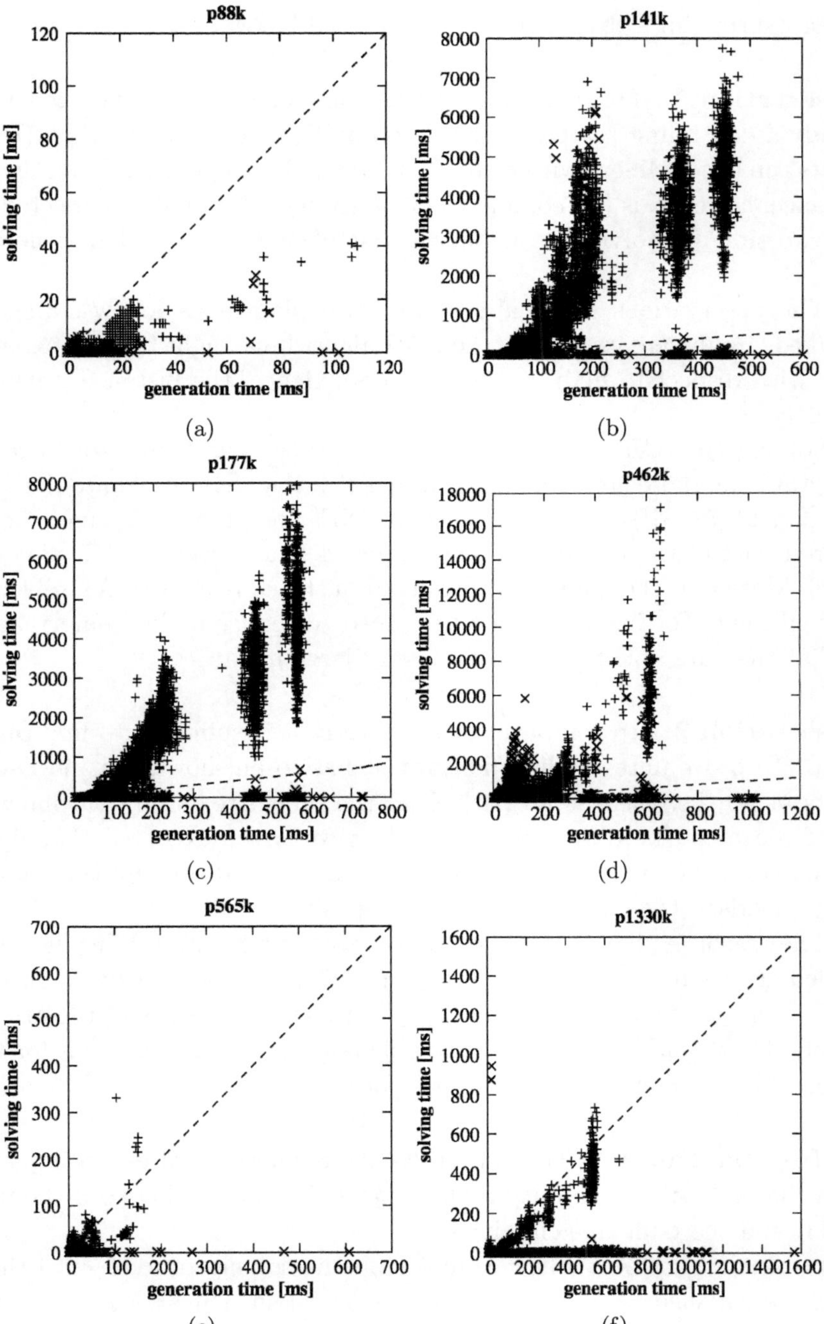

Figure 7.2: Run time comparison for individual faults

These “surprising” observations are discussed below.

Observation 1 From the theoretical point of view, instance generation is only a depth-first search algorithm on a *Directed Acyclic Graph* (DAG), i.e. its run time is linear with respect to the number of gates. Solving the SAT instance, however, is NP-complete [19]. Therefore, it would be expected that the run time for solving an instance is significantly larger than generating it.

The observations made above can be explained as follows. Since the handled circuits are very large, the DAG algorithms might become expensive (e.g. with respect to main memory access). Hence, the instance generation is costly.

Solving the SAT instance, however, is often “easy” due to its regular structure, i.e. there are many implications possible that accelerate the search (see [29, 49, 79, 82]). Additionally, most CNFs are quite small, since the considered part of the circuit (see e.g. Figure 4.1 in Section 4.1) is also quite small. Moreover, the data structures used in state-of-the-art SAT solvers are very efficient. For instance, they are tuned to reduce main memory access, so that they are able to handle even very large instances.

Observation 2 To compute a test pattern, it is sufficient to find **one** D-chain. To prove untestability, however, it has to be shown that no D-chain exists at all, i.e. there is no path from the fault site to any output where a difference can be observed. Although it would be expected that finding one solution, i.e. a test pattern, is much easier than proving that no such solution exists, the analysis shows the opposite.

The reason is that, in most cases, the source for untestability lies in the immediate environment of the fault site. This can be quickly determined due to the efficient conflict analysis incorporated in state-of-the-art SAT solvers. On the other hand, for a testable fault, much time is spent for value propagation. This is discussed in more detail below.

Testable Faults Classical ATPG algorithms stop after finding a path from the fault site to an output that shows a difference between the faulty circuit and the fault free circuit.

In contrast, a SAT solver cannot stop the solving process until the instance is satisfied. A CNF is known to be satisfied if at least one of the following statements holds:

1. All clauses are satisfied.

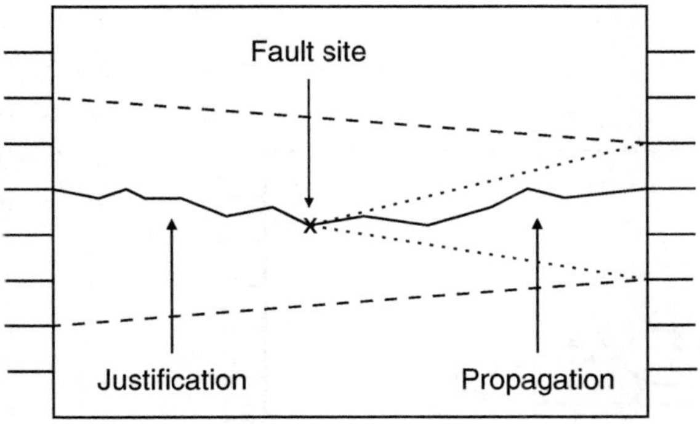

Figure 7.3: Example for a testable fault

2. All variables are assigned and no conflict occurred.

Note, the second statement implies the first one.

Modern SAT solvers like MiniSat [29] prove satisfiability using the second condition. Instead of checking every clause for satisfaction after each assignment, these solvers only check for conflicts. If each variable is assigned and no conflict occurred, the instance is satisfiable. Hence, after finding a D-chain, i.e. the fault is testable, each variable of the entire influenced circuit part has to be assigned without conflicts. Often, this is a very time consuming step.

Figure 7.3 illustrates this. Assume the fault is testable. Hence, it is possible to justify a D-value at the fault site (along the solid line from the *Primary Input* [PI] towards the fault site) and to find a D-chain (the solid line from the fault site towards the *Primary Outputs* [PO]).

As mentioned above, a classical algorithm can prove testability of the fault by assigning the variables along the solid line. (Additionally, some more variables may have to be assigned in order to justify values.) In SAT-based ATPG, however, the entire subcircuit enclosed by the dashed lines is transformed into a CNF. Afterwards, a SAT solver has to find a consistent variable assignment for this CNF. This results in considerable overhead, especially if those "negligible" areas are hard-to-solve, e.g. areas containing many symmetries and reconvergences.

Untestable Faults On the other hand, if the fault is untestable, often the conflict leading to the unsatisfiability of the SAT instance occurs quite quickly. Due to the D-variables used to encode the difference between the

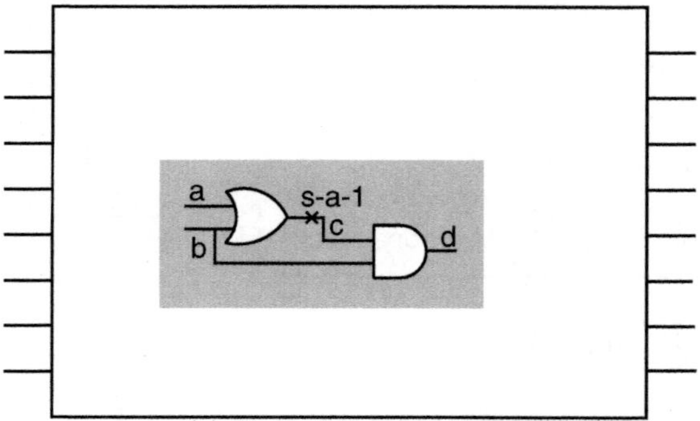

(a) Part of the circuit graph

$$\begin{aligned}
\omega_1 &: (\overline{c}_c + a_c + b_c)\\
\omega_2 &: (c_c + \overline{a}_c)\\
\omega_3 &: (c_c + \overline{b}_c)\\
\omega_4 &: (d_c + \overline{b}_c + \overline{c}_c)\\
\omega_5 &: (\overline{d}_c + b_c)\\
\omega_6 &: (\overline{d}_c + c_c)\\
\omega_7 &: (d_f + \overline{b}_c + \overline{c}_f)\\
\omega_8 &: (\overline{d}_f + b_c)\\
\omega_9 &: (\overline{d}_f + c_f)\\
\omega_{10} &: (\overline{d}_d + d_c + d_f)\\
\omega_{11} &: (\overline{d}_d + \overline{d}_c + \overline{d}_f)\\
\omega_{12} &: (\overline{c}_d + c_c + c_f)\\
\omega_{13} &: (\overline{c}_d + \overline{c}_c + \overline{c}_f)\\
\omega_{14} &: (\overline{c}_d + d_d)\\
\omega_{15} &: (c_f)\\
\omega_{16} &: (c_d)
\end{aligned}$$

(b) CNF description

Figure 7.4: Example for an untestable fault

good circuit and the faulty circuit, conflicts during propagation and justification of the fault effect occur early, and often close to the fault site.

Figure 7.4 illustrates this in detail. A part of a circuit is shown that contains two gates and a stuck-at fault on the connection between those gates. Assume that the circuit can be modeled in Boolean logic. It can be seen that this fault is untestable, since it is impossible to inject a D-value (signal b has to be set to 0) and propagate the D-value (signal b has to be set to 1) at the same time.

In Figure 7.4b the SAT instance

$$\Omega = \bigcup_{i=1}^{16} \omega_i$$

for the given ATPG problem is depicted. Due to clauses ω_{15} (injection of the fault) and ω_{16} (injection of the D-value), it is possible to propagate within this CNF until the SAT instance contains the two clauses

$$\omega_3' : (\overline{b}_c) \quad \text{and} \quad \omega_5' : (b_c),$$

i.e. the Boolean variable b has to be assigned conflicting values. Since this conflict occurred just by propagation, i.e. without making any decision, the SAT instance is unsatisfiable. Therefore, the fault is untestable.

Obviously, since the fault is untestable within this circuit part, it is untestable within the entire circuit. Let $\hat{\Omega}$ denote the CNF of the entire circuit. Then the statement

$$\Omega \subseteq \hat{\Omega}$$

holds. Since Ω is unsatisfiable $\hat{\Omega}$ is unsatisfiable as well. The SAT instance Ω is called an *unsatisfiable core* of $\hat{\Omega}$ (see Section 3.2.4). From this, the fault is proven to be untestable.

As a result, even the SAT instances of very large circuits, i.e. CNFs with large instance generation time, can be proven to be unsatisfiable within only a few propagation steps and nearly no run time.

7.2.2 Incremental Approach

In this section, based on the observations made above, an incremental solving technique is proposed to accelerate both instance generation and instance solving.

In the following, the term *fanin cone* denotes the fanin cone of a primary output.

Overview

As proposed earlier, after determining all POs belonging to the structural fanout of the fault site, the SAT instance, consisting of the transitive fanin cones of all these outputs, is built up completely. Afterwards, this CNF is solved by a SAT solver and finally, the fault is classified.

This flow is changed in the *incremental instance generation* method. Using this method, only a small portion of the circuit is converted into CNF, i.e. only a partial SAT instance is generated. This instance is solved by a SAT solver, but perhaps a fault classification cannot be given definitively. In such cases, the partial SAT instance has to be augmented.

Figure 7.5 gives an overview of the incremental instance generation algorithm. A more detailed illustration is given in Figure 7.6. It shows a circuit abstraction in four different steps of the algorithm.

The algorithm starts with an *initial CNF* only consisting of clauses modeling the injection of the fault. In this initial step, the influenced POs are also determined. This can be seen in Figure 7.6a. The POs are denoted by dotted lines.

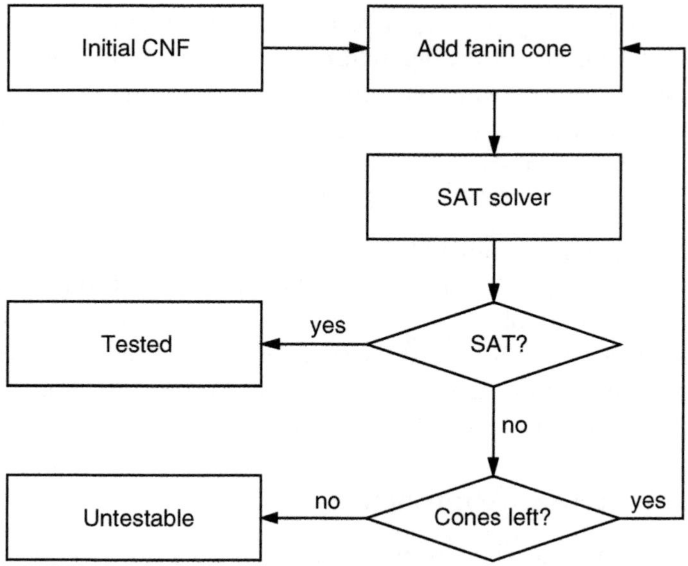

Figure 7.5: Sketch of the proposed incremental solving technique

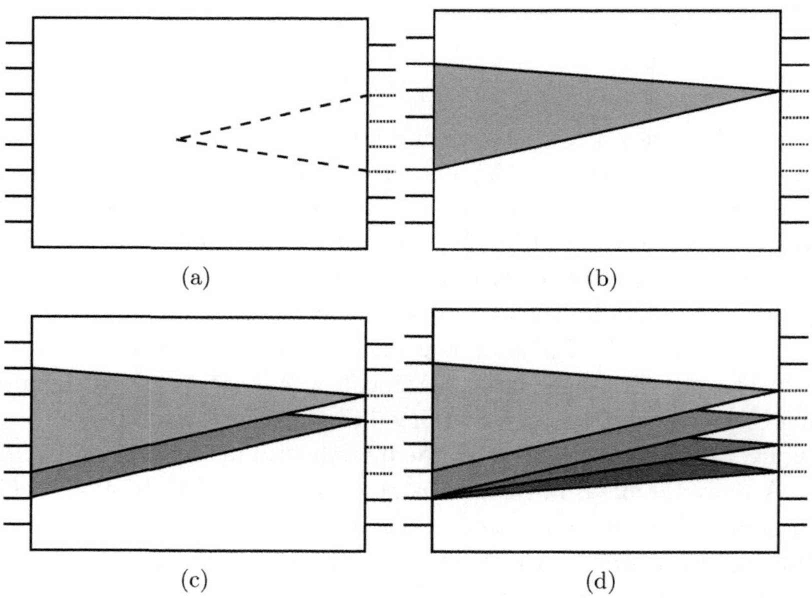

Figure 7.6: Illustration of the algorithm

Afterwards, the first PO's fanin cone is added to the current SAT instance as shown in Figure 7.6b and the resulting SAT instance is solved by the SAT solver. If it is satisfiable, the fault is tested and the algorithm terminates for this target fault.

Otherwise, no classification can be given since the fault may be observable on some other output. Therefore, the CNF is augmented by the second PO's fanin cone (see Figure 7.6c). Note, that each gate is added to the SAT instance only once, i.e. gates already contained in the CNF (due to the already traversed fanin cones) are not added a second time. This process is repeated until the fault is classified or all fanin cones have been added to the CNF (illustrated in Figure 7.6d). In the latter case, the entire SAT instance has been generated. If this CNF is unsatisfiable, the fault is untestable.

The following example further illustrates this process.

Example 19 *Figure 7.7 shows a circuit with three primary outputs (m, n and o). A stuck-at fault is modeled on signal j. Building the complete SAT instance would result in a CNF for the entire circuit, which consists of 23 variables (15 variables for the correct circuit, 4 variables for the faulty circuit and 4 D-chain variables).*

Using the incremental approach to build the instance results in a much smaller CNF. For example, the transitive fanin $\mathcal{F}(m)$ of output m consists

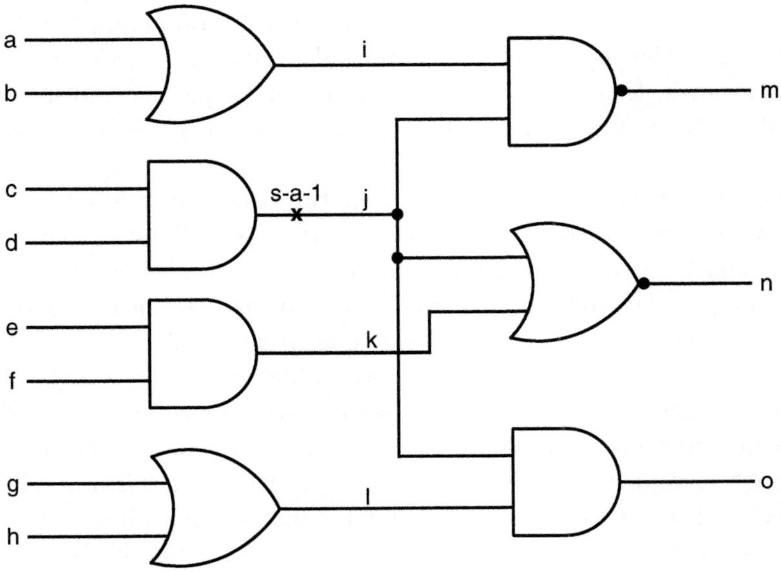

Figure 7.7: Example circuit

of gates i and j and inputs a, b, c and d. The resulting CNF has only 11 variables (7 variables for the correct circuit, 2 variables for the faulty circuit and 2 D-chain variables).

Since the fault is observable at output m, the SAT instance is satisfiable. Hence, this smaller CNF with only half of the variables needed to build up the SAT instance for the entire circuit is sufficient to classify the fault. The situation is similar for outputs n and o.

Discussion

Solving a problem incrementally is a known approach in verification. For instance, an incremental solving technique for SAT-based equivalence checking is given in [23]. Similar to the method described above, the fanin cones of the POs are taken into account incrementally. However, the incremental approach is limited to the solving step; the SAT instance is always completely generated.

In the following, improvements to the incremental instance generation method are discussed. Assume the fault is testable and a difference can be seen on the first output. Then – according to Observation 1 – the instance generation time is reduced, since only a subset of the entire circuit has to be traversed. Additionally – according to Observation 2 – requiring a smaller set of clauses to be satisfied accelerates the solving process.

Now assume the fault is testable but not observable on the first output chosen. In the worst case, the fault may only be observable on the last output. Then, the approach has to extend the respective current CNF many times. However, this worst case does not occur frequently and usually the satisfiable CNF is still smaller than the large CNF generated by the standard approach.

Moreover, in the proposed approach, the SAT solver has to be called n times. However, since $n - 1$ instances are unsatisfiable – according to Observation 2 – and learned information (in the form of conflict clauses), derived by previous solving processes, is kept during the entire classification process, the solving process can be expected to be fast.

Finally, assume that the fault is untestable. In this case, to be able to classify the fault, the entire SAT instance has to be built. Therefore, it cannot be expected to accelerate the classification process for these faults. Since each incremental step causes some overhead in the form of additional variables and clauses (details are given in the next section), it is even possible to slow down the classification process. This may happen if the number of incremental steps is too large.

However, in a typical industrial circuit, the number of testable faults exceeds the number of untestable faults significantly. Therefore, it is likely that the improvements due to incremental solving outweigh this drawback. This will be observed from the experiments presented in Section 7.3.

Implementation Details

The above description did not address the order of choice of the primary outputs. In the current implementation, the POs are ordered with respect to their distance to the fault site, i.e. short paths are preferred. The reasoning is the fact that short paths are typically easier to sensitize.

It was mentioned above that each incremental step creates some overhead with respect to the instance size. This is explained in the following.

Figure 7.8 shows three gates. Assume that signals d and e do not converge on any path to the outputs. Therefore, gates d and e are contained in different POs' fanin cones. Recall the D-chain clause given in Section 4.2. Let $h^i, 1 \leq i \leq q$ be the successors of g, then:

$$g_d \rightarrow \bigvee_{i=1}^{q} h_d^i$$

If g is on a D-chain, at least one successor of g must be on the D-chain as well. In this example, this leads to:

$$c_d \rightarrow d_d \vee e_d$$

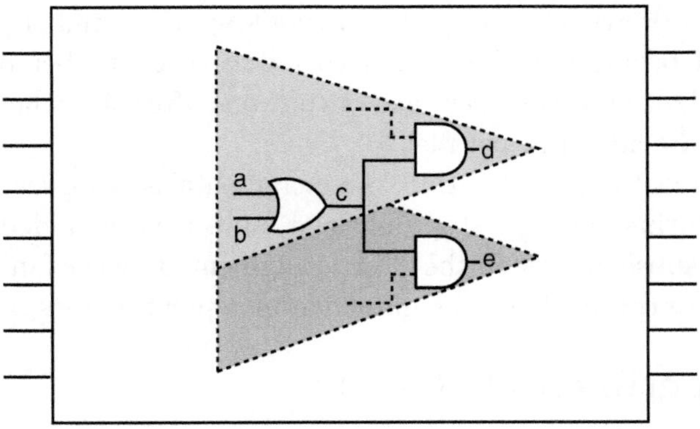

Figure 7.8: Example of the overhead during different incremental solving steps

Thus, building the entire instance at one time, the clause

$$\omega = (\overline{c}_d + d_d + e_d)$$

has to be added to the SAT instance, in order to propagate the difference towards the outputs.

Now, assume the SAT instance is built using the proposed incremental approach. When adding fanin cone $\mathcal{F}(d)$ the clause

$$\omega_1 = (\overline{c}_d + d_d)$$

is added and afterwards when adding fanin cone $\mathcal{F}(e)$, the clause

$$\omega_2 = (\overline{c}_d + d_d + e_d)$$

is added.

It can be easily seen that clause ω_1 covers clause ω_2. Therefore, clause ω_1 has to be removed from the CNF. If ω_1 was kept in the SAT instance, only a solution which sensitizes a path over d would be legal. However, since for efficiency in modern SAT solvers like MiniSat removing clauses from the database is not supported, a literal λ is added to each of these D-chain clauses in order to activate or deactivate them by *incremental assumptions* [29].

Assigning the corresponding variable v_λ of λ in such a way that λ is satisfied, all clauses containing λ are satisfied (deactivated). Assigning v_λ in such a way that λ is unsatisfied, all clauses containing λ are activated.

To summarize, each incremental step creates overhead in the form of one variable v_λ (to realize the D-chain clause activation and deactivation) and a few clauses. Since, especially for very large instances, this overhead can become a drawback during the solving step, the number of incremental steps should be limited. An option to reduce the number of incremental steps would be to not consider each output one after the other, but to build groups that are added in parallel.

In the current implementation, the incremental solving scheme is realized as follows: during the first step only one fanin cone is added to the CNF. During each subsequent step the CNF is augmented by the fanin cones of one fourth of all outputs. Hence, at most five incremental steps are performed.

7.3 Experimental Results

In this section, experimental results for the proposed techniques are presented. All experiments were carried out on an Intel Xeon (3 GHz, 32,768 MByte RAM, GNU/Linux). More information about benchmarking and about industrial circuits can be found in Section 2.5. The SAT solver MiniSat [29] v1.14 was used. A difference between the experimental results in Sections 7.3.1 and 7.3.2 can be observed. This is due to the use of different versions of the overall ATPG framework.

7.3.1 Hybrid Logic

In this section, experimental results of the hybrid logic approach, presented in Section 7.1, are given. In Table 7.2, further information about the industrial circuits is presented. Column *Targets* shows the number of target faults, while column *Gates* presents the total number of gates in the circuit. The absolute and the relative number of gates that have to be handled with four-valued logic is shown in column *Gates 4v* and in column *% 4v*, respectively. This overview is only made for industrial circuits since the ITC'99 benchmarks can be modeled completely using Boolean logic.

Three circuits contain only Boolean elements. On the other hand, one third of the elements of circuit p177k must be handled with four-valued logic.

Table 7.3 gives information about the SAT instance sizes with respect to the number of variables (column *Variables*) and the number of clauses (column *Clauses*) using the four-valued logic representation for every gate and using the hybrid logic representation, respectively. Furthermore, the average clause length, i.e. the average number of literals a clause consists of, is compared between the two approaches in columns *∅Len.*

Due to the typically very small number of gates that can assume non-Boolean values, the average sizes of the instances are significantly reduced. This applies to the number of clauses as well as to the number of

Table 7.2: Circuit statistics

Circuit	Targets	Gates	Gates 4v	% 4v
p44k	64,105	41,625	0	0
p49k	142,461	48,592	0	0
p80k	197,834	76,837	0	0
p88k	147,048	83,610	7,236	8.65
p99k	162,019	90,712	1,571	1.73
p177k	268,176	140,516	47,512	33.81
p462k	673,465	417,974	73,085	17.49
p565k	1,025,273	530,942	112,652	21.22
p1330k	1,510,574	950,783	109,690	11.54

Table 7.3: Average instance sizes – hybrid logic usage

	Four-valued logic			Hybrid logic		
Circuit	Variables	Clauses	⌀Len	Variables	Clauses	⌀Len
b14	7,587	32,930	3.03	4,230	11,753	2.34
b15	9,801	45,964	3.25	5,588	15,917	2.38
b17	8,830	39,663	3.14	4,711	13,139	2.35
b18	8,617	37,316	3.08	4,569	12,575	2.33
b20	10,637	46,605	3.04	5,691	15,872	2.34
b21	10,727	46,888	3.03	5,907	16,502	2.34
b22	10,593	46,457	3.04	5,757	16,063	2.35
p44k	60,446	209,001	3.14	30,708	74,435	2.30
p49k	204,130	1,092,143	3.43	101,038	316,294	2.54
p80k	7,483	22,697	3.13	4,211	9,471	2.37
p88k	5,047	15,676	2.88	2,672	6,359	2.34
p99k	5,301	16,139	2.83	2,854	6,504	2.34
p177k	69,908	222,516	2.87	47,087	134,933	2.50
p462k	7,336	22,399	2.77	5,175	14,784	2.39
p565k	4,663	15,638	2.82	2,661	6,956	2.37
p1330k	20,580	62,929	2.81	18,163	55,475	2.54

variables. The average length of the clauses of the SAT instances is reduced, too. Note, that the size of the circuit does not correlate to the average size of the SAT instances.

The results in terms of run time and number of aborted faults are shown in Table 7.4. Again, column *Four-valued logic* gives the number of aborted faults (*Aborts*) and the run time (*Time*) of the approach using four-valued logic. The number of aborted faults and the run time of the hybrid approach can be found in column *Hybrid logic*. Time is given either in CPU minutes (*min*) or in CPU hours (*h*).

Test generation for one target fault is aborted if the instance cannot be solved after seven MiniSat restarts. The results show that the number of aborted faults of the hybrid approach is – in comparison to modeling all signals with four-valued logic – significantly reduced in almost all cases.

Due to the heuristic nature of SAT solvers, it can also happen that the run times are longer, even though the SAT instance is more compact (see p1330k). But as the experiments show, this rarely happens. Typically, smaller instances also directly result in run time savings. In the experiments, improvements of nearly a factor of 8 could be observed (p177k).

Table 7.4: Experimental results

	Four-valued logic		Hybrid logic	
Circuit	Aborts	Time	Aborts	Time
b14	0	1:00 min	0	0:19 min
b15	0	1:16 min	0	0:24 min
b17	0	4:36 min	0	2:22 min
b18	0	27:33 min	0	22:30 min
b20	0	2:30 min	0	0:56 min
b21	0	2:41 min	0	0:59 min
b22	0	3:49 min	0	1:35 min
p44k	0	2:18 h	0	26:01 m
p49k	*Timeout*		77	1:43 h
p80k	1	42:58 min	0	9:43 min
p88k	0	11:41 min	0	9:33 min
p99k	0	8:41 min	0	6:50 min
p177k	941	10:28 h	0	1:19 h
p462k	129	3:31 h	6	2:16 h
p565k	0	2:23 h	0	2:23 h
p1330k	1	4:58 h	1	5:05 h

Moreover, circuit p49k can be solved using hybrid logic, while the previous SAT-based approach using only four-valued logic failed within the given overall run time limit (20 CPU hours). The experimental results clearly show that using hybrid logic instead of only four-valued logic improves the performance of SAT-based ATPG significantly with respect to run time and number of aborted faults and, thus, yields a more robust ATPG process.

7.3.2 Incremental Instance Generation

In this section, the experimental results of the incremental instance generation approach are given.

In Figure 7.9, the run time analysis made in Section 7.2.1 is repeated using the proposed method. It can be seen that most of the testable faults (denoted by '+') can be classified with a significant speed-up of both instance generation and solving. As predicted in Section 7.2.2, the proposed method has only small influence on untestable faults (denoted by '×').

In Table 7.5, the average CNF sizes, i.e. the number of variables (column *Variables*) and the number of clauses (column *Clauses*), using traditional SAT-based ATPG and using the proposed incremental approach are given.

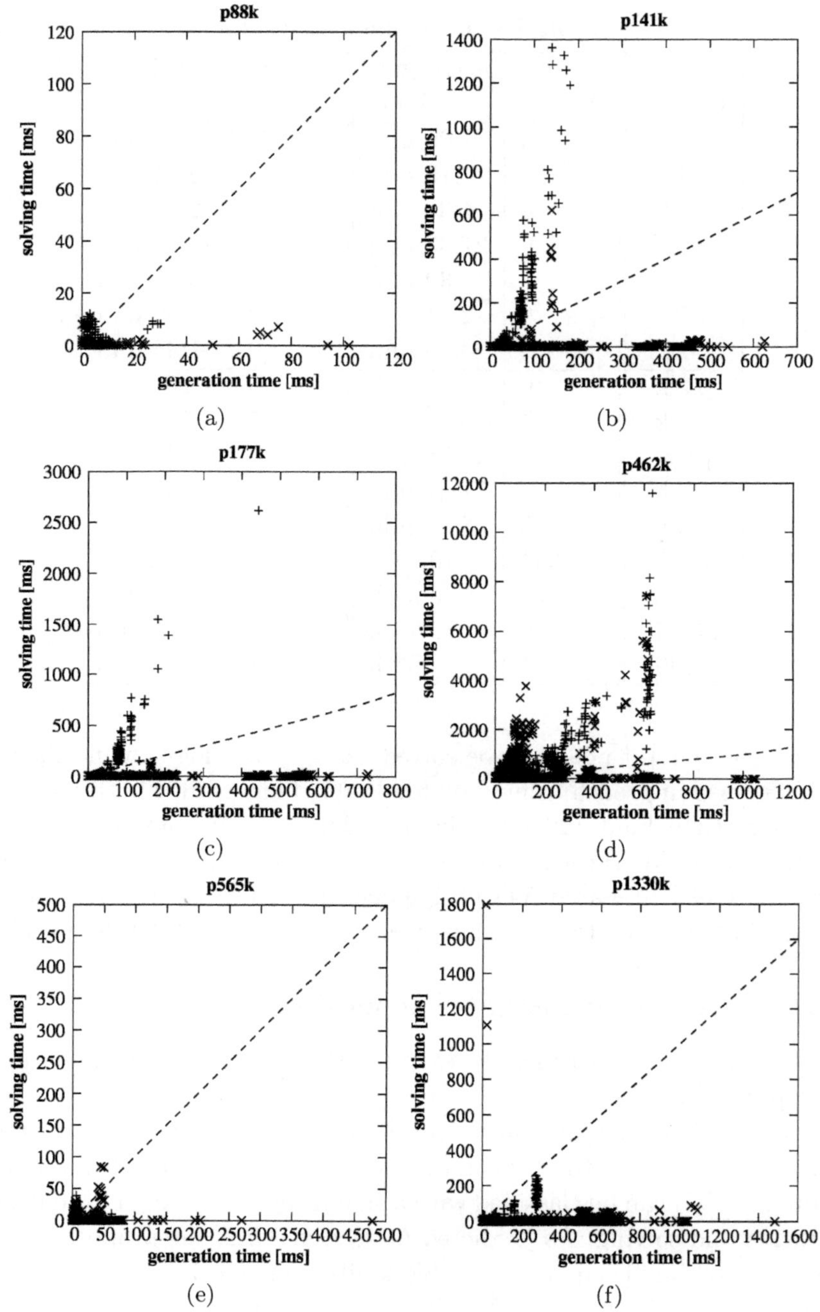

Figure 7.9: Run time comparison for individual targets based on the incremental approach

Table 7.5: Average CNF sizes

	Traditional SAT		Incremental approach	
Circuit	Variables	Clauses	Variables	Clauses
b17	6,424	16,693	3,613	9,046
b18	6,134	15,667	3,262	7,918
b20	7,383	19,433	2,854	7,028
b21	7,452	19,627	2,906	7,160
b22	7,420	19,533	2,667	6,511
p44k	29,819	72,767	21,011	49,269
p77k	544	1,374	378	934
p80k	4,312	9,930	1,369	2,848
p88k	2,366	5,570	1,244	2,744
p99k	2,589	5,955	1,367	2,992
p141k	33,521	95,672	18,782	53,249
p177k	37,775	109,659	21,386	61,807
p456k	6,727	18,611	5,772	16,257
p462k	4,365	12,530	3,790	10,779
p565k	1,681	4,316	1,326	3,445
p1330k	16,704	52,338	15,510	48,871
p2787k	16,911	56,483	16,679	56,609
p3327k	34,377	75,002	27,929	59,981
p3852k	20,622	47,205	14,557	33,253

Both approaches use hybrid logic as presented in this chapter. In the case of the proposed incremental method, the numbers given refer to the SAT instance size after the fault has been classified. In both approaches, only clauses added during the circuit-to-CNF conversion (see Section 4.1) are given, i.e. no conflict clauses are considered.

It can be seen that using the proposed method results in smaller average CNF sizes than using the traditional approach. For circuit p80k, the average number of clauses is reduced to less than one third. In one case (circuit p2787k), the number of clauses increases slightly. This can be explained by that circuit's unusually high number of untestable faults. For all other benchmarks, significant reductions can be observed.

Table 7.6 gives an overview of the overall run times using traditional SAT-based ATPG and the proposed method. For each run, the run time (columns *Time*) and the number of aborts (columns *Aborts*) is shown. An abort occurs after seven MiniSat restarts. Time is given either in CPU minutes (*min*) or in CPU hours (*h*).

Table 7.6: Run times for the ATPG process

	Traditional SAT		Incremental approach	
Circuit	Aborts	Time	Aborts	Time
b17	0	2:51 min	0	1:29 min
b18	0	9:07 min	0	4:12 min
b20	0	2:18 min	0	0:46 min
b21	0	2:22 min	0	0:49 min
b22	0	2:59 min	0	0:57 min
p44k	0	49:11 min	0	15:18 min
p77k	0	0:18 min	0	0:12 min
p80k	0	6:30 min	0	1:01 min
p88k	0	2:19 min	0	1:15 min
p99k	2	1:35 min	1	1:00 min
p141k	1	3:02 h	0	22:17 min
p177k	0	2:35 h	0	24:32 min
p456k	194	39:03 min	182	31:33 min
p462k	11	1:09 h	9	42:38 min
p565k	0	6:35 min	0	5:42 min
p1330k	0	1:02 h	0	54:22 min
p2787k	1,628	14:55 h	1,433	12:37 h
p3327k	1,833	48:38 h	838	18:38 h
p3852k	1,484	17:32 h	604	8:25 h

The experimental results show that the use of the proposed incremental method results in a significant speed-up of up to a factor of 8 (circuit p141k). The number of aborted faults is reduced as well. This method scales especially well for the large industrial circuits (p3327k, p3852k) which exhibit many aborted faults.

7.4 Summary

In this chapter, two improvements to the circuit-to-CNF conversion procedure have been proposed. Both techniques lead to smaller SAT instances and also to reduced run time of the ATPG process. The number of aborted faults is further reduced.

First, the use of a hybrid logic for industrial circuits has been introduced. Since those circuits contain non-Boolean gates, a four-valued logic has to be used. However, a large portion of a circuit is Boolean and can be modeled with Boolean logic.

During a preprocessing step, the circuit parts that can be represented with Boolean logic are determined. By using four-valued logic only where necessary and Boolean logic where possible, the SAT instance sizes can be decreased.

Second, a detailed analysis of state-of-the-art SAT-based ATPG algorithms with respect to their run time for single faults has been provided. It has been shown that, firstly, instance generation often needs more run time than solving the instance and, secondly, it is often more complex to prove testability than to prove untestability.

Based on these observations, an incremental SAT instance generation technique that accelerates both instance generation and solving the instance has been proposed. The experimental results confirm that the overall run time of the ATPG computation and the number of aborted faults can be significantly reduced.

It can be concluded that integrating the new techniques in the existing SAT-based ATPG approach leads to a more robust ATPG process that can cope with very large industrial circuits.

Chapter 8

Branching Strategies

In this chapter,[1] branching strategies for SAT-based ATPG algorithms are presented. In Section 8.1, the concepts of the standard SAT solver decision heuristics are reviewed, while in Section 8.2, a combination of decision strategies from a structure-based algorithm and from a SAT solver is discussed. Standard SAT solver decision strategies as well as a structural branching strategy are experimentally evaluated in Section 8.3. A summary is presented in the last section.

8.1 Standard Heuristics of SAT Solvers

As described in Chapter 3, a SAT solver traverses the search space by a backtracking scheme. Although BCP and conflict analysis have greatly improved the speed of SAT solvers, the variable selection strategy remains crucial to achieving an efficient traversal of the search space. No general way to choose the best variable is known, as the decision about satisfiability of a given CNF formula is NP-complete [19].

Therefore, SAT solvers have sophisticated heuristics to select variables as explained in Section 3.2.3. Usually, the heuristic accumulates some statistics about the CNF formula dynamically during the execution of the SAT solver. Then, this data is used as the basis for decisions. This leads to a trade-off between the quality of a decision and the overhead needed to update the statistics. Also, the quality of a given heuristic often depends on the problem domain. The default variable selection strategy applied by modern SAT solvers like zChaff [82] and MiniSat [29] is the quite robust VSIDS strategy

[1] A preliminary version of this chapter has been published in [91].

R. Drechsler et al., *Test Pattern Generation using Boolean Proof Engines*,

which was explained in Section 3.2.3. As a potential drawback, VSIDS does not use any structural knowledge of the circuit.

Decisions based on variable selection also occur in classical test pattern generation where structural methods or approximate measures, e.g. SCOAP [51], are usually employed to determine a good choice for the next variable selection. In contrast to the dynamic procedure of VSIDS, structural methods usually are static. In the following, it is explained how structural heuristics and the VSIDS heuristic can be combined.

8.2 Decision Strategies

Making decisions only on primary inputs was the improvement of PODEM [48] over the D-algorithm. Any other internal value can be implied from the primary inputs. This yields a reduction of the search space and motivates applying the same strategy to SAT-based test pattern generation. For SAT solving, this is done by restricting the variable selection of the SAT solver to those variables corresponding to primary inputs or state bits of the circuit. The VSIDS strategy is applied to these variables to benefit from the feedback of conflict analysis and the current position in the search space. Figure 8.1 depicts this strategy. Only the variables in the dotted oval are considered for selection.

Restricting the variable selection to fanout gates only was first proposed in FAN [44]. Again, the idea is to restrict the search space while achieving a large number of implications from a single decision. Conflicts resulting from a decision are often due to a small region within the circuit. However, in our experiments, the application of a fanout-based decision heuristic has

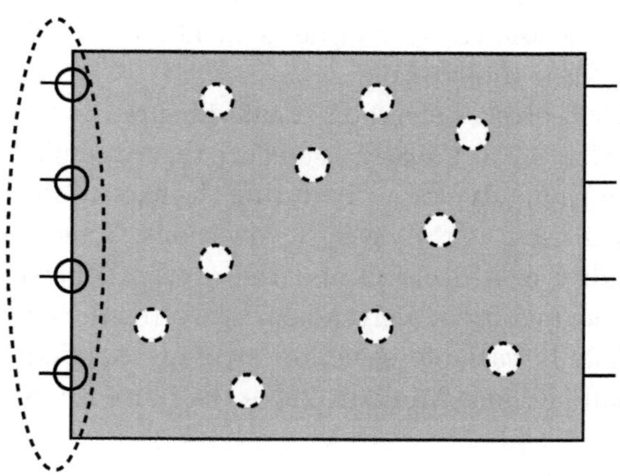

Figure 8.1: Illustration of Branching-on-Input heuristic and VSIDS heuristic

not provided any advantage in combination with VSIDS. The benefits of a fanout-based heuristic are subsumed by the VSIDS heuristic.

The following decision schemes are therefore proposed and experimentally evaluated:

- VSIDS – Decision variables are determined by the standard VSIDS heuristic.
- BoI (Branch-on-Inputs) – Decisions are carried out only on the inputs of the circuit. The VSIDS heuristic is only applied to the inputs.

The difference between the BoI heuristic and the VSIDS heuristic is illustrated in Figure 8.1. Black bordered circles mark decisions that are determined by the BoI heuristic. In contrast, circles bordered with dotted lines show additional decision points of the VSIDS heuristic. As the experiments in the next section show, no heuristic is in general superior. Both heuristics have advantages for some circuits. Therefore, two sequential approaches are further proposed:

- BoI–VSIDS – A combination of BoI and VSIDS with BoI performed first. If BoI does not lead to a solution in a given time or restart interval, VSIDS is activated and decisions are carried out on any variable.
- VSIDS–BoI – The converse of the combination above. First, the VSIDS heuristic is applied. Decision variables are selected among all variables. If this does not yield a test pattern, BoI is activated and decision variables are restricted to the primary inputs.

8.3 Experimental Results

In this section, the experimental results of the different decision strategies are presented and discussed. The experiments were performed on an Intel Xeon (3 GHz, 32,768 MByte RAM, GNU/Linux). Benchmark circuits from the ITC'99 benchmarks suite as well as industrial circuits provided from NXP Semiconductors were used. Statistical information about the industrial circuits can be found in Section 2.5. MiniSat [29] v1.14 was used as the SAT solver.

The results are shown in Table 8.1. The first column gives the circuit's name, whereas the other columns show the results of the corresponding decision strategy. For the sequential approaches BoI–VSIDS and VSIDS–BoI, the heuristic was changed after eight MiniSat restarts, the general timeout was 12 restarts.

Table 8.1: Results for different decision strategies

	VSIDS		BoI		BoI–VSIDS		VSIDS–BoI	
Circ	Ab.	Time	Ab.	Time	Ab.	Time	Ab.	Time
b14	0	**0:33 min**	1	0:47 min	1	0:40 min	0	**0:33 min**
b15	0	**0:45 min**	0	1:13 min	0	1:12 min	0	**0:45 min**
b17	0	**2:17 min**	0	3:08 min	0	3:10 min	0	**2:17 min**
b18	0	7:44 min	2	10:24 min	1	9:48 min	0	**7:41 min**
p44k	0	52:32 min	0	26:44 min	0	**26:33 min**	0	51:58 min
p49k	1,223	12:17 h	209	11:15 h	240	3:44 h	244	**2:44 h**
p77k	0	**0:06 min**	0	**0:06 min**	0	**0:06 min**	0	**0:06 min**
p80k	0	3:45 min	0	3:56 min	0	3:56 min	0	**3:43 min**
p88k	0	**1:45 min**	0	2:27 min	0	2:27 min	0	**1:45 min**
p99k	0	**1:04 min**	1	1:52 min	5	1:40 min	1	**1:04 min**
p177k	0	1:10 h	0	49:51 min	0	**49:40 min**	0	1:10 h
p456k	7	23:04 min	34	58:22 min	116	42:02 min	45	**20:12 min**
p462k	0	**45:45 min**	108	4:47 h	83	2:46 h	10	46:44 min
p565k	0	**6:50 min**	30	8:50 min	0	7:16 min	0	**6:50 min**
p1330k	0	50:16 min	0	51:51 min	0	51:31 min	0	**49:58 min**

Comparing VSIDS and BoI, VSIDS is in most cases faster and has fewer aborts. For circuit p462k, VSIDS is faster by a factor of more than 6. VSIDS has only two circuits with aborted faults – the least among the four approaches. Nonetheless, the number of aborted faults for circuit p49k is the largest. In contrast, the run time of BoI for p44k is nearly half of that of VSIDS and the number of aborts is reduced by a factor of nearly 6 for p49k.

Concerning the sequential approaches, the following observations can be made. In general, BoI–VSIDS is comparable to BoI and VSIDS–BoI is comparable to VSIDS. This can be explained by the number of decision scheme changes, given in Table 8.2. As described above, if no solution is found after eight MiniSat restarts, the second decision scheme is applied. The numbers given in this table, represent the number of times the first restart limit was reached and the second scheme was applied for the circuit. In most cases, the number of changes is very low. But it can be seen that decision scheme changes occur more frequently in BoI–VSIDS than in VSIDS–BoI. This is an indicator that solutions are usually found faster using VSIDS–BoI.

Comparing BoI and VSIDS with BoI–VSIDS and VSIDS–BoI, respectively, the run time is further reduced by the sequential approaches but more aborts are produced; except for p49k, a special case which can be considered as a "hard-to-test" circuit. Here, the VSIDS–BoI approach reduces the number of aborts and the run time significantly compared to VSIDS.

Table 8.2: Number of decision scheme changes for the sequential approaches

Circ	Bol–VSIDS	VSIDS–Bol
b14	1	0
b15	0	0
b17	4	0
b18	18	0
p44k	0	0
p49k	708	2,099
p77k	0	0
p80k	0	0
p88k	0	0
p99k	10	1
p177k	1	0
p456k	484	224
p462k	862	17
p565k	30	0
p1330k	3	0

These results show that VSIDS is a robust heuristic, producing only a few aborts. However, for hard-to-test circuits, BoI is more robust. BoI–VSIDS produces too many aborts and is faster than VSIDS–BoI only in two cases. Usually, VSIDS–BoI is faster than VSIDS, but suffers from a slightly increased number of aborts (if not considering p49k). But – more importantly – VSIDS–BoI is also applicable for hard-to-test circuits, e.g. p49k, and is thus quite robust. Hence, in the following chapters, VSIDS–BoI is used as the standard heuristic in PASSAT.

8.4 Summary

Branching heuristics are a key feature of a SAT solver. However, the standard heuristics are based on statistics and do not use structural information. As the experimental results presented here show, standard SAT heuristics are fast but produce many aborts in "hard-to-test" circuits. The proposed structure-based heuristic is much slower in most cases but produces only a few aborts in hard-to-test circuits. The experiments also show that a combination of both heuristics provides a robust alternative to the standard SAT heuristic.

Chapter 9

Integration into Industrial Flow

Thus far, the SAT-based ATPG approach PASSAT has been considered as a standalone ATPG tool. In this chapter,[1] how to integrate it into an industrial flow is discussed.

In the first section, the problem is motivated and a brief overview of the ATPG flow in an industrial environment is given. An effective integration of a SAT-based ATPG engine into an industrial environment is described in Section 9.2. First, observations made during initial experiments are presented. The advantages and drawbacks of (classical) ATPG engines are compared with those of a SAT-based algorithm. Afterwards – motivated by those observations – a concrete combination of the two approaches is proposed.

In Section 9.3, a weakness of SAT-based ATPG is targeted. Modern SAT solvers prove satisfiability by finding a *complete* variable assignment. As a result, test patterns derived by SAT-based ATPG contain a large number of specified bits. This is disadvantageous since the test patterns cannot be compacted very well. An approach to decrease the number of specified bits is proposed. A post-processing step computes sufficient variable assignments to the gates using structural information. As a result, the number of don't cares can be increased significantly. The proposed methods are experimentally evaluated in Section 9.4. Finally, this chapter is summarized in Section 9.5.

[1]Parts of Section 9.2 have been published in [26], while a preliminary version of Section 9.3 has been published in [30].

R. Drechsler et al., *Test Pattern Generation using Boolean Proof Engines*,

9.1 Industrial Environment

In principle, it is sufficient to iterate over all faults with respect to the given fault model and generate a test pattern for each of them. However, in an industrial environment, this is not sufficient to achieve a robust system. Typically an ATPG framework consists of several interacting steps and engines (see Section 2.3). In the following, the overall flow in an industrial system will be briefly reviewed to explain the problems that occur during the integration of a SAT-based engine. For a more detailed presentation on ATPG systems in general, please refer to [59].

The major steps of an ATPG flow are shown in Figure 9.1. This is a more detailed view than the one presented in Section 2.3. The inputs for the system are the circuit and the fault model to be considered. Here, the SAFM is assumed. Two main steps are carried out: the pre-identification phase to classify faults and the compaction phase to generate a small test set.

The goal during pre-identification is the classification of faults. Here, three engines are used. First, random test pattern generation is applied to filter out "easy-to-detect" faults. For the remaining faults, a fast deterministic test pattern generation is carried out. Finally, deterministic test pattern generation with increased resources is applied to classify "hard-to-classify" faults.

As a result, four classes of faults are generated: untestable faults, easy-to-detect faults, testable faults and non-classified aborted faults. Untestable faults are not further considered. Only the remaining testable faults are further treated in the compaction step. Note, that in the pre-identification step, all generated test patterns are discarded. This phase only determines the fault classes.

In the compaction step, test patterns that detect as many faults as possible are generated. Mostly, faults that are not easy to detect are targeted. Easy-to-detect faults are likely to be detected without targeting them explicitly, i.e. by fault simulation. This phase uses the information gathered in the pre-identification step. As a result, a small test set is generated. Small test sets reduce test time during post-production test. Furthermore, a small test set needs a small amount of memory on the tester itself.

The main step considered here is the deterministic fault detection applied in the pre-identification phase. In this phase, it is important to classify as many faults as possible. Only faults classified as testable are considered during the *compact pattern generation*. Aborted faults are considered during fault simulation as well. Therefore, untestable faults that were not classified, i.e. aborted faults, are an overhead in the compaction step. Testable faults that were not classified may not be detected by the *compact TPG*.

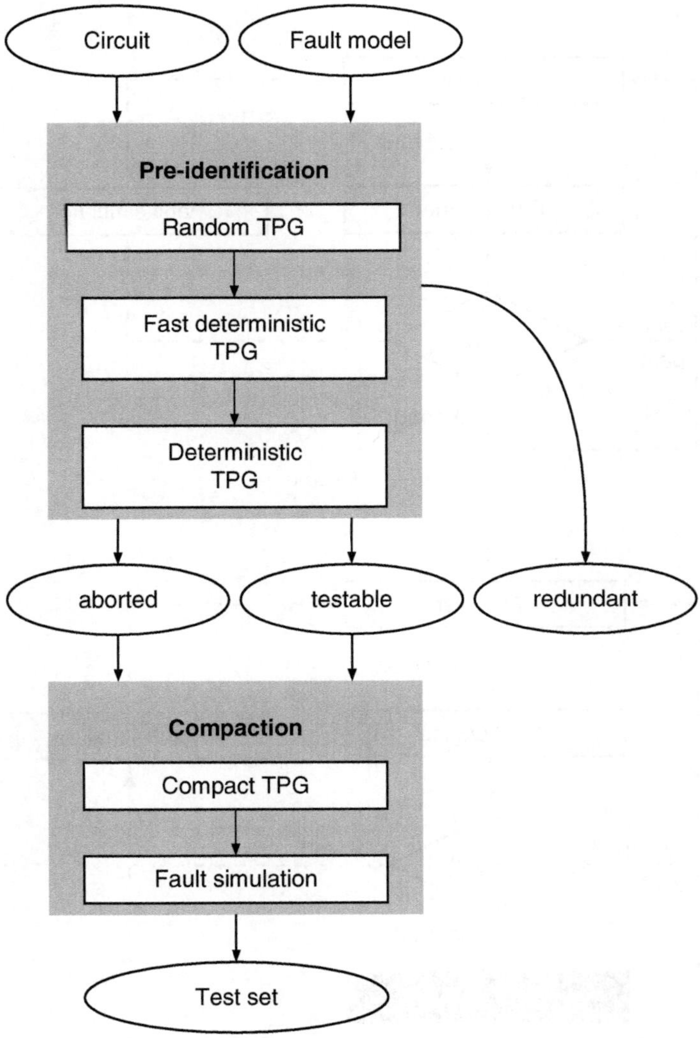

Figure 9.1: ATPG flow in an industrial environment

The details of deterministic pattern generation are shown in Figure 9.2a. Usually, a highly optimized structural ATPG engine and a fault simulator are applied for deterministic pattern generation. The ATPG framework of NXP Semiconductors applies a highly optimized FAN-based engine including learning techniques.

To begin, the FAN-based engine is used to classify a given target fault. If a test pattern is produced by this engine, additional faults may be detected

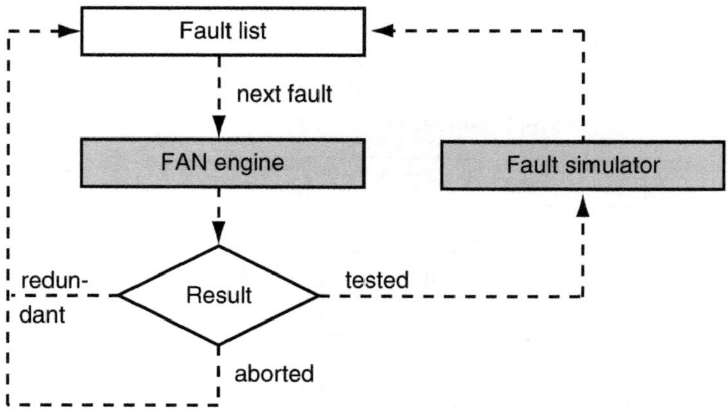

(a) Classic

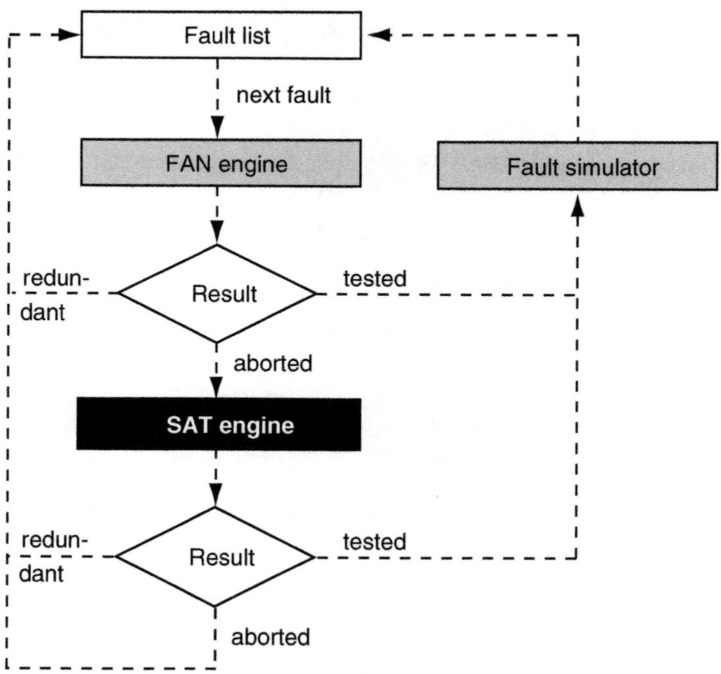

(b) SAT integration

Figure 9.2: Deterministic test pattern generation

by the pattern besides the fault initially targeted. Therefore, a fault simulator is used to determine all faults detected by this test pattern. This consumes additional computation time, but often speeds up the overall process because many other faults can be removed from the fault list.

9.2 Integration of SAT-Based ATPG

When integrating a SAT-based ATPG engine into an industrial environment, the goal is to improve the overall performance of the system. Two aspects have to be considered: the run time and the number of classified faults. The run time should be decreased. The number of faults that are classified by the system should be increased, i.e. the number of aborted fault classifications due to resource limits should be decreased.

The following concentrates on integrating the SAT-based engine into the pre-identification phase since, as explained in the previous section, reducing the number of aborted faults in this step is beneficial for the succeeding compaction step. Furthermore, a high fault coverage, i.e. as few aborts as possible, is important to ensure that the manufactured chip is correct.

The underlying problem is to determine how to interlace the engines. The integration of random pattern generation and deterministic pattern generation is typically done by applying random pattern generation in a fast preprocessing step. Similar to a FAN-based approach, the SAT-based engine needs time to generate the problem instance before solving can be started. This overhead is much smaller when random pattern generation is applied, i.e. many faults have already been classified as tested. Therefore random pattern generation precedes the SAT-based engine in the flow. In the following we address how to effectively use a SAT-based engine in the standard flow between the FAN-based engine and fault simulation.

The "classic" deterministic flow is illustrated in Figure 9.2a. The FAN-based engine takes a target fault from the fault list. If the fault is testable, the fault simulator is called for the generated test pattern. All faults detected by this test pattern are removed from the fault list. Then, the next fault is targeted by the FAN-based engine.

Numerous observations help to set up the framework for the integration of the SAT approach:

- There are faults that are easily classified by FAN while SAT needs a long run time and vice versa. This behavior is not predictable before performing test pattern generation for a particular fault.

- A large number of faults can be classified efficiently using FAN.
- Often the SAT-based engine efficiently classifies untestable faults as well as faults that are hard for FAN, i.e. those faults where FAN needs long run times or aborts due to pre-defined resource limits (as explained in Section 7.2.1).
- The FAN-based algorithm runs directly on the circuit structure which is already available in the system.
- The SAT-based algorithm converts the problem into a CNF before starting the SAT solver. Therefore, a larger overhead per fault is needed compared to FAN.
- A SAT solver determines values for all inputs that are contained in the transitive fanin of those outputs where the fault may be observed. This makes merging of multiple test-patterns during compaction difficult. (A possible approach to overcome this drawback is shown in the next section.)

These observations led to the conclusion that the SAT-based engine should be used to target those faults that cannot be classified by the FAN-algorithm within a short time interval. This avoids overhead for initializing the SAT-based engine on faults that are easy to classify by FAN. Then, the SAT-based approach may classify hard faults which, in turn, helps to remove other faults from the fault list.

Given a fault list, this leads to the framework shown in Figure 9.2b. The FAN-based engine is started at first with a short time interval. If a test pattern is generated, fault simulation is carried out as usual and may identify additional faults as being testable by the same test pattern.

If the fault is untestable, the classification process can stop immediately. In other words, if the FAN-based algorithm is able to classify the fault, the SAT-based engine is not started. The ATPG process continues with the next fault.

However, if the FAN-based engine is not able to classify the fault, i.e. the given resources like time limit or backtrack limit are exceeded, the SAT-based engine is applied in order to classify the fault in a second step. The experiments, presented in Section 9.4.1, show that this combined approach classifies more faults with almost no overhead for the additional runs of the SAT-based engine.

9.3 Test Pattern Compactness

In this section, techniques to reduce the size of a test pattern in SAT-based ATPG are shown. SAT-based ATPG algorithms have been shown to be effective even for large industrial circuits. But a weakness of this method is the large number of specified bits in the computed test patterns.

During their search for a solution of a problem, state-of-the-art SAT solvers, e.g. zChaff [82] or MiniSat [29], prove either the unsatisfiability by showing that no solution for the given formula exists or the satisfiability by computing a satisfying assignment for the formula. The stopping criterion of the latter case is the complete assignment of all variables.

More formally, a solution of a Boolean formula $f(x_1, \ldots, x_n)$ is found if, and only if,

$$\forall x_i \mid 1 \leq i \leq n: \; x_i \in \{0, 1\}$$

and no contradiction exists.

From this solution, the test pattern is directly determined by the assignment of the input variables. Due to the complete Boolean assignment, all bits of the test pattern of the considered part of the circuit have a specified value. That means they are either 0 or 1, but not X (*don't care*).

In contrast, classical ATPG algorithms such as FAN [44] or SOCRATES [88] assign X-values to signals during their search process and, as a result, directly generate test patterns with a smaller number of specified bits.

In industrial practice, it is important, that computed test patterns have a large number of unspecified bits. This is required to apply techniques like test compaction and test compression effectively.

In the following, strategies are presented that reduce the number of specified bits in test patterns computed by SAT-based ATPG. Two post-processing strategies that result in more compact test patterns are introduced. In Section 9.3.1, the exploitation of structural properties about the observability of the fault effect is presented, while Section 9.3.2 applies local don't cares. A similar technique for structural ATPG was proposed in [81].

9.3.1 Observability at Outputs

For a given stuck-at fault, the fault must be justified at the faulty line and then propagated towards the outputs, so that the fault effect is observable at at least one output.

During the generation of the SAT instance, it is not known along which paths the fault effect will be propagated. Therefore, all possible paths and their transitive fanin cone have to be included. As a result, the test pattern

Algorithm 3 Pseudo code of the post-processor

```
1: testpattern t = X
2: set<input> s
3: list l
4: for all output o do
5:   if observable(o) then
6:     l.push(o)
7:     while !l.empty() do
8:       gate g = l.first_element()
9:       if g == INPUT then
10:        s.add(g)
11:      else
12:        l.add(g.all_predecessors())
13:      end if
14:      l.remove(g)
15:    end while
16:    for all input i ∈ s do
17:      t.set_computed_bit(i)
18:    end for
19:    break
20:  end if
21: end for
```

is over-specified, i.e. many bits of the test pattern are not required to justify or propagate the fault effect.

To reduce the number of specified bits in the test pattern t, a post-processor is applied after calculating the solution. Algorithm 3 shows the pseudo-code of the post-processor.

First, all bits of the test pattern are set to X (line 1). Then, the inputs s of the transitive fanin cone of output o ($\mathcal{F}(o)$) at which the fault effect is observable are identified by backtracing (line 7–15). The assignments of all inputs in s are extracted and the corresponding bits in the test pattern are set to the required values (line 17). Because it is sufficient that the fault effect can be observed at one output, it is sufficient to apply this procedure only once. Therefore, the complexity of this post-processing step is $\mathcal{O}(n)$, where n denotes the number of elements in the circuit for a randomly chosen o.

However, because it is possible that the fault effect can be observed at more than one output, the number of specified bits in the test pattern depends on the chosen output. To find the output with the smallest number

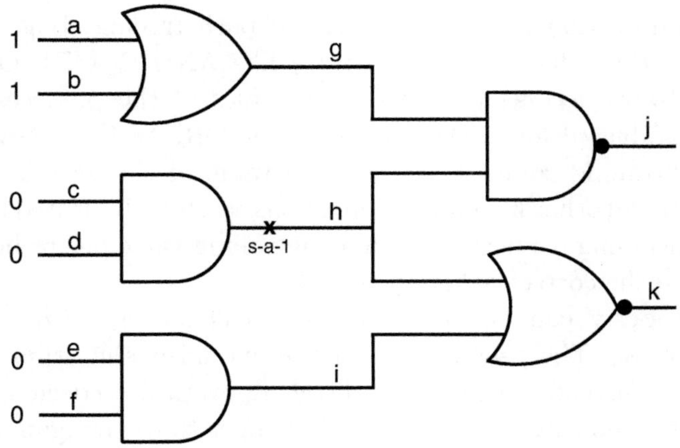

Figure 9.3: Example circuit

of specified bits, the procedure must be executed for each output o at which the fault effect is observable. In this case, the complexity of the procedure is $\mathcal{O}(k \cdot n)$, where k denotes the number of outputs in the output cone.

Example 20 *Consider the circuit given in Figure 9.3. A s-a-1 fault is to be tested at line h. A test pattern found by the classical SAT approach is given to the left of the inputs: $a = 1, b = 1, c = 0, d = 0, e = 0, f = 0$.*

The fault effect can be observed at both outputs. Choosing output j, the post-processor backtraces to the inputs a, b, c, d and sets the corresponding bits in the test pattern to the specified value. The bits for the inputs e, f remain X. Choosing output k, however, results in specified bits for inputs c, d, e, f and don't care bits for a, b.

In this example, the reduction of the specified bits is 33%.

9.3.2 Applying Local Don't Cares

In this section, a procedure which exploits the knowledge about local don't cares is introduced. This procedure can be combined with the technique presented in the previous section.

In the procedure described above, the inputs are identified by backtracing without considering internal assignments. Consequently, all inputs on which the output o structurally depends are represented by specified bits in the test pattern. But not all considered internal signals in $\mathcal{F}(o)$ are typically necessary for detecting the fault.

A procedure that is similar to critical path tracing [1] is applied. For determining the value of a basic gate g like AND, NAND, OR or NOR, it is not always necessary to know all values of the predecessors. If the controlling value (0 for AND, NAND; 1 for OR, NOR) is assumed by at least one incoming connection, then the value of the gate is determined. Consequently all other incoming connections can be substituted by X-values. Only one incoming connection with a controlling value has to be considered to guarantee the correct value of the gate.

This property can be exploited when calculating $\mathcal{F}(o)$ under a specific assignment. The pseudo-code of the extended algorithm is shown in Algorithm 4. Instead of directly backtracing over all predecessors of a gate of the circuit, the gate is analyzed with respect to the assignment.

Algorithm 4 Pseudo code of the post-processor applying local don't cares

```
testpattern t = X
set<input> s
list l
for all output o do
   if observable(o) then
      l.push(o)
      while !l.empty() do
         gate g = l.first_element()
         if g == INPUT then
            s.add(g)
         else if on_d_chain(g) then
            l.add(g.all_predecessors())
         else if contr_in_val(g) then
            l.add(g.pred_with_contr_val())
         else
            l.add(g.all_predecessors())
         end if
         l.remove(g)
      end while
      for all input i ∈ s do
         t.set_computed_bit(i)
      end for
      break
   end if
end for
```

If the gate is located on a D-chain, i.e. the fault effect is propagated through this gate, all predecessors must be considered (line 11–12). This is due to the requirement that all side-inputs of the D-chain must be set to a non-controlling value to propagate the fault effect.

If the assignment of at least one incoming connection of the considered gate is the controlling value (line 13), then only the corresponding predecessor has to be considered. The other predecessors are not addressed and can therefore be treated as X. Note, that this is not an exact method. When there is more than one controlling value, the choice is based on heuristics.

In all other cases, all predecessors must be considered to ensure the correct value. Finally, the bits of all inputs which have been considered during backtracing are set to the value assigned by the test pattern.

Analogous to the procedure presented in Section 9.3.1, the number of specified bits depends on the chosen output. To determine the output with the smallest number of specified bits for the given assignment, the same procedure is applied repeatedly.

Example 21 *Consider again the circuit in Figure 9.3. Choosing k as the observed output results in considering all predecessors of k, because k is on a D-chain, i.e. h and i have to be considered. Both gates h, i assume the controlling value on their outputs. Therefore, those incoming connections have to be considered for backtracing with the controlling value of the gate, i.e. the value* 0.

Consequently, it is sufficient to consider only one predecessor of h and i, respectively. This results in only two specified bits (one of $\{c, d\}$ and one of $\{e, f\}$) which are needed to detect the fault instead of all six.

Here, the reduction of the specified bits is 66%.

9.4 Experimental Results

In this section, experimental results are given. First, results for the integration of a SAT-based ATPG algorithm into an industrial environment are shown. Second, results for the application of the test pattern compaction methods are given. Further details about benchmarking and statistical information about the industrial circuits can be found in Section 2.5.

9.4.1 Integration

The proposed integration of a SAT-based engine into the industrial environment was applied to the ATPG framework of NXP Semiconductors in a prototypical manner. The SAT-based ATPG algorithm PASSAT is used. The underlying SAT solver is MiniSat [29] v1.14. As in SOCRATES [88] and HANNIBAL [65], the main ATPG engine is a highly optimized FAN-algorithm. PASSAT has been integrated into the system as explained in Section 9.2.

The resource limits of the FAN-based algorithm were set to the default settings. These parameters have been determined on a large range of circuits. The restart interval was used as the timeout for the SAT-based engine. Test generation for a single fault was aborted after 12 MiniSat restarts. The focus of the discussion is the methodology of the integration of a SAT-based approach into an industrial environment. Optimizing parameter settings is not the focus of the discussion.

As a result, a loose coupling between the engines is achieved. All faults that would not be classified in the classical flow are targeted by the SAT-based engine afterwards. Consequently, more faults can be classified in total. The SAT-based engine runs in two steps: first, the CNF is generated, then, the fault is classified; afterwards the CNF is completely dropped. Reusing parts of the CNF and learned information for other faults is not considered, because the SAT-based engine is only applied to aborted classifications that are "randomly" distributed. Therefore, identifying structural overlapping of CNF instances is difficult.

For all other settings in the flow, e.g. the number of random patterns to be simulated, the default parameters of the industrial framework were used. As benchmarks, the industrial circuits provided by NXP Semiconductors and the publicly available ITC'99 benchmarks are used. Most of the experiments were run on an AMD Athlon XP 3500+ (2.2 GHz, 1,024 MByte RAM, GNU/Linux). For the larger circuits p2787k, p3327k and p3852k, experiments were carried out on an AMD Opteron (2.8 GHz, 32,768 MByte RAM, GNU/Linux) – due to the increased memory requirements.

Four different approaches are compared in the following:

- *SAT*: Using the SAT-based engine only, i.e. PASSAT
- *FAN*: Using the FAN-based engine with default parameters
- *FAN (long)*: Using the FAN-based engine with drastically increased resources, i.e. increased backtracking limit and increased time limit
- *FAN + SAT*: Using the combined SAT and FAN approach as explained in Section 9.2

Table 9.1: Experimental results for the classification

	SAT		FAN		FAN (long)		FAN+SAT	
Circuit	Ab.	Time	Ab.	Time	Ab.	Time	Ab.	Time
b14	0	0:19 min	107	0:11 min	7	1:42 min	0	0:12 min
b15	0	0:24 min	619	0:11 min	318	26:25 min	0	0:18 min
b17	0	2:22 min	1,382	1:41 min	622	56:54 min	0	1:58 min
b18	0	22:30 min	740	19:16 min	270	41:40 min	0	20:34 min
b20	0	0:56 min	225	0:35 min	42	7:46 min	0	0:44 min
b21	0	0:59 min	198	0:39 min	43	6:48 min	0	0:43 min
b22	0	1:35 min	284	1:07 min	52	9:34 min	0	1:14 min
p44k	0	26:01 min	12	4:58 min	0	4:59 min	0	5:55 min
p49k	77	1:43 h	3,770	2:06 h	162	2:38 h	74	1:55 h
p57k	2	3:51 min	225	1:34 min	142	9:45 min	2	1:44 min
p80k	0	9:43 min	218	34:55 min	21	39:13 min	0	39:38 min
p88k	0	9:33 min	195	9:13 min	38	12:40 min	0	10:27 min
p99k	0	6:50 min	1,398	6:02 min	512	1:16 h	0	7:25 min
p141k	0	1:27 h	276	1:58 min	69	3:04 min	0	2:53 min
p177k	0	1:19 h	270	16:06 min	47	20:03 min	0	19:03 min
p456k	10	31:35 min	6,919	18:34 min	3,296	3:05 h	11	29:58 min
p462k	6	2:16 h	1,383	1:34 h	423	2:07 h	0	1:51 h
p565k	0	2:23 h	1,391	2:21 h	85	2:47 h	0	2:47 h
p1330k	1	5:05 h	889	4:15 h	144	4:28 h	0	5:00 h
p2787k	0	15:28 h	215,206	1:46 h	147,565	20:41 h	0	11:54 h
p3327k	112	45:29 h	32,483	5:15 h	15,001	13:33 h	101	14:13 h
p3852k	88	33:54 h	34,158	8:56 h	19,171	16:39 h	81	9:17 h

A combination of *FAN (long)* and *SAT* is not considered here. Those faults that cannot be classified by *FAN* but by *FAN (long)* can typically be classified by the SAT engine much faster. The run time of a combination of *FAN (long)* and *SAT* is not acceptable.

Table 9.1 shows the results of the classification phase. For each of the four approaches, the number of aborted fault classifications (column *Ab.*) and the run time (column *Time*) are given. Time is given either in CPU minutes (*min*) or CPU hours (*h*). The number of aborted faults is most critical because these may not be targeted adequately in the compaction step as explained in Section 9.1.

PASSAT is able to classify all ITC'99 benchmarks and most of the industrial circuits completely. If the FAN-based engine is used, no circuit can be classified completely. On the other hand, using this approach, the run time is reduced significantly. Except for circuits p49k and p80k, the FAN algorithm is always faster than the SAT-based engine. This is due to the overhead for the generation of the SAT instance.

If the FAN-based engine is allowed to use increased resources, i.e. the FAN (long) approach, the classification time increases. However, this decreases the number of aborted faults in many cases – often by one order of magnitude. However, even with increased resources, only one circuit (p44k) could be fully classified.

Regarding the FAN+SAT approach, the run time is similar to that of the ordinary FAN approach for the small circuits. For the large circuits, the run time increases. However, the number of aborts decreases significantly using the combined approach. For circuit p3327k, the number of aborts actually dropped from 215,206 and 147,565, respectively, to 0. This is possible because many easy-to-detect faults were quickly classified by FAN while difficult faults were classified by the SAT-based engine. In that case, the classification success outweighs the instance generation overhead.

In summary, the combined approach is able to fully classify 17 out of 22 circuits, while the resources needed remain similar to that of a classical approach. Therefore, the integration of the SAT-based ATPG engine yields a fast and robust ATPG framework.

9.4.2 Test Pattern Compactness

In this section, the experimental results of the post-processor for improving the test pattern compactness are presented. All experiments were carried out on an AMD64 3500+ (2.2 GHz, 4,096 MByte RAM, GNU/Linux). As SAT solver, MiniSat [29] v1.14 was used.

The general test procedure is as follows. For a given fault, a test pattern is computed. Afterwards, the post-processor is started to reduce the number of specified bits. Additionally, a fault simulator is started to check whether the test pattern finds additional faults. In the following, the run time overhead of the algorithms is discussed, followed by an analysis of the quality.

The run time overhead of the post-processor is shown in Table 9.2. Note, that the exact run time of the post-processor was not measured, but the difference between the total ATPG run time with and without post processor is reported. Due to the use of a fault simulator, both ATPG runs can behave totally differently, because other faults are targeted.

In column *Post*, the run time overhead of the approach presented in Section 9.3 is shown. Column *Post ext.* gives the run time overhead for the approach applying local don't cares which was presented in Section 9.3.2.

Table 9.2: Results – run time overhead

Circuit	Time	Post Def	Post Min	Post ext. Def	Post ext. Min
p44k	48:13 min	+0:11 min	−0:15 min	+2:42 min	+2:56 min
p49k	2:14 h	−0:01 h	−0:01 h	−0:01 h	+0:04 h
p77k	0:30 min	−0:01 min	−0:01 min	−0:01 min	−0:01 min
p80k	11:54 min	+17:30 min	+17:29 min	+18:14 min	+18:16 min
p88k	12:36 min	+0:23 min	+0:29 min	+0:15 min	+0:20 min
p99k	9:35 min	−0:05 min	−0:02 min	+0:05 min	+0:10 min
p177k	1:40 h	−0:01 h	−0:01 h	+0:11 h	+0:13 h
p462k	2:58 h	−0:01 h	−0:02 h	+0:01 h	+0:01 h
p565k	2:38 h	+0:01 h	0:00 h	+0:01 h	+0:01 h
p1330k	6:01 h	+0:06 h	+0:09 h	−0:05 h	−0:05 h

In the columns entitled *Def*, the run time overhead for a randomly chosen output is provided. Columns entitled *Min* report the run time overhead for determining the output with the smallest number of specified bits. Note, that a '+' symbol means, that the run time is higher with the post-processor and a '−' means that the run time of the approach without the post-processor is higher.

Studying the results of Table 9.2, it can be observed, that the additional use of the post-processor only results in small run time overhead. This is due to the linear complexity of the algorithm. Only in the case of p80k, the run time is more than doubled. This can be explained by the corresponding increased number of test generator calls, i.e. the fault simulator finds a smaller number of additional faults detected by the test patterns. In all other cases, this significant increase of the calls cannot be observed.

Comparing the configurations *Post Def* and *Post Min*, it can be seen that using *Post Min* results more often in (slightly) smaller run time, although more calculations have to be done. Again, this is due to the use of a fault simulator, i.e. different test patterns cause a different set of targeted faults. This cannot be observed using *Post Ext. Def* and *Post Ext. Min*. Here, the run time of *Post Ext. Def* always remains smaller (or equal) compared to *Post Ext. Min*.

Compared to the ATPG run without the post-processor, the overhead using the post-processor is in most – but not all – cases negligible with the notable exception of p80k. The number of unclassified faults remains almost stable and are therefore not reported here.

Table 9.3: Specified bits – post-processing

	Classic		Post			Post min		
Circ	%Bits	#Pat	%Bits	%Cla	#Pat	%Bits	%Cla	#Pat
p44k	70.01	5,946	59.31	76.56	5,542	59.31	76.56	5,542
p49k	47.38	379	17.86	37.82	373	17.85	37.80	376
p77k	1.94	123	0.59	31.27	121	0.59	31.83	125
p80k	10.05	4,025	4.99	49.71	10,694	4.99	49.71	10,694
p88k	4.38	5,757	2.95	67.05	5,890	2.95	67.05	5,890
p99k	5.48	3,300	4.23	77.15	3,285	4.23	77.15	3,285
p177k	23.24	3,755	7.97	34.29	3,890	7.93	34.12	3,846
p462k	0.85	9,316	0.30	35.31	9,223	0.30	35.42	9,198
p565k	0.24	8,638	0.16	66.80	8,664	0.16	67.26	8,715
p1330k	0.26	12,151	0.17	69.81	12,477	0.17	69.81	12,477

Table 9.4: Specified bits – post-processing applying local don't cares

	Classic		Post ext.			Post min ext.		
Circ	%Bits	#Pat	%Bits	%Cla	#Pat	%Bits	%Cla	#Pat
p44k	70.01	5,946	7.59	9.72	6,149	7.59	9.72	6,149
p49k	47.38	379	16.76	35.48	368	16.68	35.32	377
p77k	1.94	123	0.49	26.51	118	0.49	26.51	118
p80k	10.05	4,025	3.17	31.64	12,915	3.17	31.64	10,985
p88k	4.38	5,757	1.15	26.23	5,752	1.15	26.23	5,752
p99k	5.48	3,300	1.52	27.86	3,354	1.52	27.86	3,354
p177k	23.24	3,755	0.69	2.99	4,086	0.70	3.00	4,113
p462k	0.85	9,316	0.13	15.92	9,254	0.14	15.96	9,235
p565k	0.24	8,638	0.09	39.79	8,695	0.09	39.83	8,729
p1330k	0.26	12,151	0.04	16.59	11,967	0.04	16.59	11,967

In Tables 9.3 and 9.4, results concerning the average number of specified bits are presented for the approaches with and without applying local don't cares, respectively. In column *%Bits*, the average percentage of specified bits of the approaches is provided (in *Classic*, this is the number of inputs included in the SAT instance). Column *%Cla* gives the percentage of specified bits in relation to the *Classic* configuration without post-processor. Finally, column *#Pat* denotes the number of generated test patterns. Note, that the test patterns are not compacted, i.e. the number corresponds to the number of test generator calls.

The use of a post-processor without applying local don't cares reduces the specified bits significantly. The results show a reduction of up to 69%.

It can be noticed, that there are only slight differences between the configurations *Post* and *Post min*. Although in *Post min*, the output with the smallest number of specified bits is considered, the total number of specified bits is not always smaller. This is again due to the use of a fault simulator.

Applying the post-processor considering local don't cares results in an even smaller percentage of specified bits in test patterns. In the worst case, the number of specified bits is reduced to only 35% of the specified bits of the *Classic* approach, while in the best case (p177k) only 3% remain.

The experiments show, that the presented post-processor is able to reduce the number of specified bits drastically.

9.5 Summary

The integration of a SAT-based ATPG engine into an industrial environment has been shown. The reason for applying the SAT-based engine as a second deterministic ATPG step to classify aborted faults has been explained in detail. Experimental results have shown the improved robustness achieved by the combination of classical ATPG algorithms with a SAT-based approach. Even on large industrial circuits that are hard to test the proposed combined approach performs better than classical engines alone and reduces the number of aborted faults dramatically.

Moreover, the problem of over-specified test patterns has been addressed. By calculating sufficient variable assignments during a post-processing step, the number of specified bits has been reduced drastically. With nearly no overhead in run time, SAT-based ATPG algorithms are able to generate compact test patterns which are well suited for techniques like test compaction and compression.

Chapter 10

Delay Faults

So far, the SAT techniques introduced have been applied to stuck-at fault test pattern generation. However, due to the shrinking feature sizes of modern circuits and their increased speed, testing of delay faults becomes more and more important. As an extension of the introduced SAT techniques, SAT-based ATPG for delay faults is presented in this chapter.[1]

The most common delay fault models are the *Transition Delay Fault Model* (TDFM) and the *Path Delay Fault Model* (PDFM). The PDFM is more accurate but due to the exponential growth in the number of paths in today's circuits, testing of all paths is not feasible. Typically, only a small number of paths, i.e. the critical paths, are considered for test generation. The TDFM is not as accurate as the PDFM, but provides good fault coverage. Therefore, it is often applied in practice.

SAT-based ATPG for *Transition Delay Faults* (TDF) is explained in Section 10.1, whereas SAT-based ATPG for *Path Delay Faults* (PDF) is described in detail in Section 10.2. One of the main differences between test generation for stuck-at faults and test generation for delay faults is that for delay faults, modeling of two consecutive time frames is needed. Furthermore, tests for delay faults may have different quality. Generally, test generation for delay faults is described for sequential circuits with standard scan design using the launch-on-capture scheme [63].

Similar to the SAFM in industrial circuits, a multiple-valued logic is needed for PDF test generation. Section 10.3 discusses the influence of the chosen Boolean encoding on the performance and shows how to determine an efficient encoding.

[1]Parts of Section 10.1 have been published in [33], whereas preliminary versions of Sections 10.2 and 10.3 have been published in [31] and [32], respectively.

R. Drechsler et al., *Test Pattern Generation using Boolean Proof Engines*,

Then an alternative approach that uses incremental SAT to generate tests of high quality is considered in Section 10.4.

Experimental results for both types of delay faults and for the different Boolean encodings are provided in Section 10.5. A summary of SAT-based ATPG for delay faults is given in Section 10.6.

10.1 Transition Delay

In this section, the modeling of TDFs for SAT-based ATPG is discussed in detail. As shown in [46], TDFs can be modeled by injecting stuck-at faults with the circuit modeled in two consecutive time frames, commonly denoted as the initial timeframe t_1 and the final timeframe t_2. In the initial time frame t_1, the faulty line must be set to the initial value of the test – 0 in case of a rising transition and 1 in case of a falling transition. Then for a rising (falling) TDF, a s-a-0 (s-a-1) fault is injected at the faulty line in the final time frame t_2 to guarantee the detection of the faulty value, i.e. a delay, in t_2 and its propagation to an output.

The two consecutive time frames t_1, t_2 are modeled using the time frame expansion (or iterative logic array representation). For this, the circuit C is duplicated. The original circuit C_1 represents t_1 and the duplicated circuit C_2 represents t_2. Further, in contrast to the SAFM, state elements, i.e. flip-flops, must be modeled. Only the initial value can be scanned into a state element in a standard scan design. The final value of a state element is calculated by the combinational logic during t_1 (launch-on-capture). Therefore, state elements are modeled by connections between the state elements in C_1 and their counterparts in C_2. This procedure is also called *unrolling*.

To apply a SAT solver to the problem, the unrolled circuit C_t must be transformed into a CNF derived from the following equation:

$$\Phi_{C_t} = \Phi_{C_1} \cdot \Phi_{C_2} \cdot \Phi_{seq}$$

The CNF for C_1 is represented by Φ_{C_1}, whereas Φ_{C_2} is the CNF for C_2. The term Φ_{seq} describes the sequential behavior of C_t, i.e. the modeling of state elements, in a standard scan design. By omitting Φ_{seq}, the CNF would represent a combinational circuit or an enhanced scan design. Note, that the CNF of the circuit is not derived directly from Boolean logic, but from the Boolean encoding of the multiple-valued logic presented in Chapter 6.

The stuck-at fault is injected at the faulty line in t_2 in the same way as described in Chapter 4. The overall constraints Φ_{TDF} for the TDF modeling contain the fixed faulty line in t_1 and the CNF for the faulty cone needed for fault injection. A test for a TDF can then be created using the following CNF Φ_{Test}:

$$\Phi_{Test} = \Phi_{C_t} \cdot \Phi_{TDF}$$

Example 22 *Figure 10.1 shows an example circuit in its original form with a falling TDF at line e. The unrolled circuit is presented in Figure 10.2. After duplicating the circuit, the pseudo primary output g is the input of the corresponding pseudo primary input* a_2 *in the duplicated circuit. To initialize the test in the initial time frame, line* e_1 *is fixed to 1, whereas a s-a-1 fault is injected at* e_2 *to propagate the fault in the final time frame to an output. The example circuit with the injected stuck-at fault (see Chapter 4) is shown in Figure 10.3.*

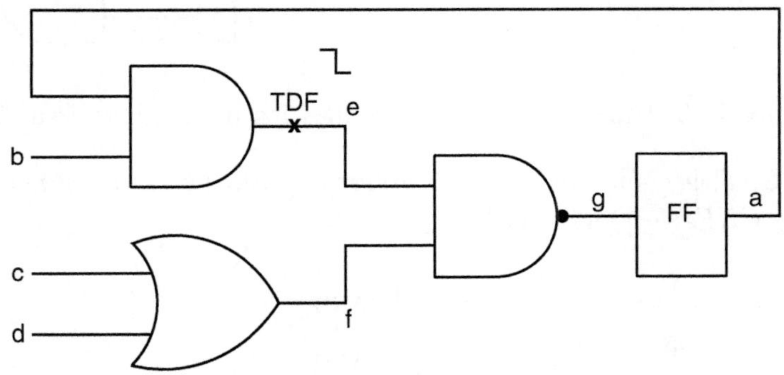

Figure 10.1: Example circuit with TDF

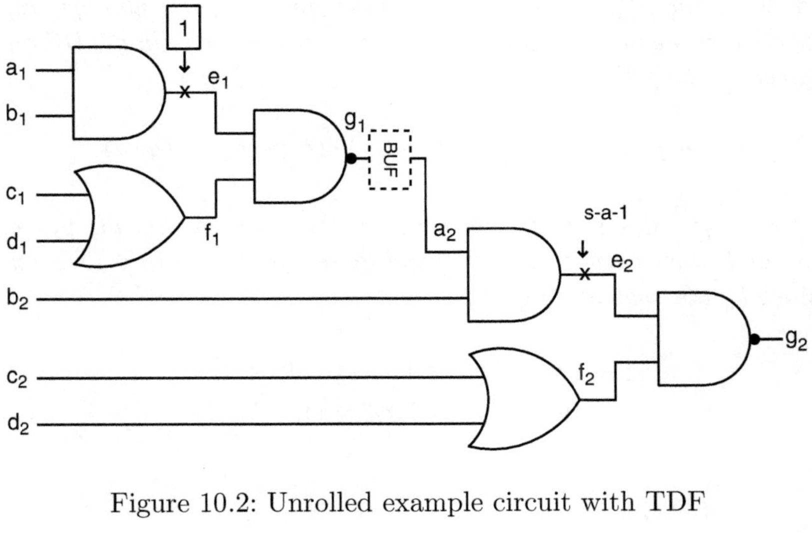

Figure 10.2: Unrolled example circuit with TDF

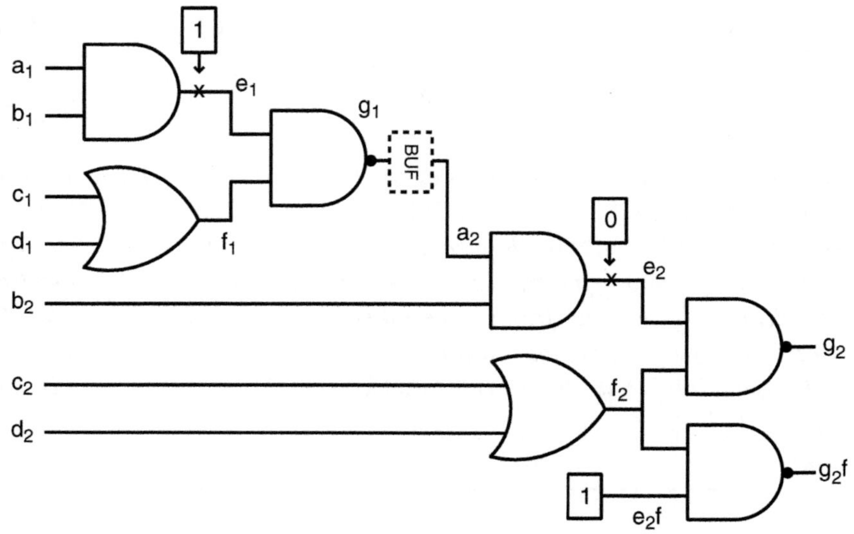

Figure 10.3: Unrolled example circuit with injected stuck-at fault

The CNF Φ_{C_t} for the unrolled circuit is given by the conjunction of the following CNFs Φ_{C_1}, Φ_{C_2} and Φ_{seq}:

$$\Phi_{C_1} = \Phi^{e_1}_{\text{AND}} \cdot \Phi^{f_1}_{\text{OR}} \cdot \Phi^{g_1}_{\text{NAND}}$$
$$\Phi_{C_2} = \Phi^{e_2}_{\text{AND}} \cdot \Phi^{f_2}_{\text{OR}} \cdot \Phi^{g_2}_{\text{NAND}}$$
$$\Phi_{seq} = (g_1 \vee \overline{a}_2) \cdot (\overline{g}_1 \vee a_2)$$

where the term $\Phi^{\text{signal}}_{\text{gatetype}}$ denotes the CNF for the particular gatetype and the particular signal. The constraints for modeling the falling TDF on line e are given in the following equation:

$$\Phi_{TDF} = \Phi^{g_2 f}_{\text{NAND}} \cdot \Phi^{g_2 D}_{\text{D}} \cdot (e_1) \cdot (\overline{e}_2) \cdot (e_2 f) \cdot (g_2 D)$$

The term $\Phi^{g_2 D}_{\text{D}}$ describes the encoding of the D-chain, i.e. the propagation of the fault to an output. A corresponding test for the falling TDF on line e obtained by the evaluation of Φ_{Test} is:

$$v_1 = \{a_1 = 1, b_1 = 1, c_1 = 1, d_1 = 0\}$$
$$v_2 = \{b_2 = 1, c_2 = 1, d_2 = 1\}$$

Note, that in practice an additional constraint must be added. The values of a primary input must often be equivalent in both time frames. This is due to the test equipment, where it is hard to change the test value on the primary inputs at speed during the test application. This constraint is incorporated by the following CNF Φ_{eq}:

$$\begin{aligned}\Phi_{eq} = &(\overline{b}_1 \vee b_2) \cdot (b_1 \vee \overline{b}_2) \cdot (\overline{c}_1 \vee c_2) \cdot (c_1 \vee \overline{c}_2) \\ &\cdot (\overline{d}_1 \vee d_2) \cdot (d_1 \vee \overline{d}_2)\end{aligned}$$

The test presented above is therefore invalid, because d has no valid assignment in practice. A valid test is:

$$\begin{aligned}v_1 &= \{a_1 = 1, b_1 = 1, c_1 = 1, d_1 = 0\} \\ v_2 &= \{b_2 = 1, c_2 = 1, d_2 = 0\}\end{aligned}$$

10.2 Path Delay

In this section, test generation for PDFs is described in detail. First, the generation of non-robust tests is explained in Section 10.2.1. The generation of robust tests is described in Section 10.2.2 and the handling of further constraints, coming from the industrial application, are considered in Section 10.2.3. An incremental formulation of static values that exploits the fact that non-robust and robust test generation is often performed sequentially is proposed in Section 10.4.

10.2.1 Non-robust Tests

As described in Section 2.2.2, two time frames are needed for a non-robust test. Therefore, two Boolean variables x_1^s, x_2^s are assigned to each connection; each of describing the value of s in the corresponding time frame. The CNF for each gate is duplicated using the respective variables resulting in the CNF Φ_{C_1} for the initial time frame and Φ_{C_2} for the final time frame. To guarantee the correct sequential behavior, additional constraints Φ_{seq} describe the functionality of the flip-flops as described for TDFs in Section 10.1. These constraints guarantee the equivalence of the value of a pseudo primary output in t_1 and the value of the corresponding pseudo primary input in t_2. The CNF representation $\Phi_{C_{NR}}$ of the *unrolled* circuit for non-robust PDF test generation can be derived in the same way as for the TDF test described in Section 10.1:

$$\Phi_{C_{NR}} = \Phi_{C_1} \cdot \Phi_{C_2} \cdot \Phi_{seq}$$

Table 10.1: Off-path input constraints (replicated from Table 2.1)

	Rising rob.	Falling rob.	Non-rob.
AND/NAND	$X1$	$S1$	$X1$
OR/NOR	$S0$	$X0$	$X0$

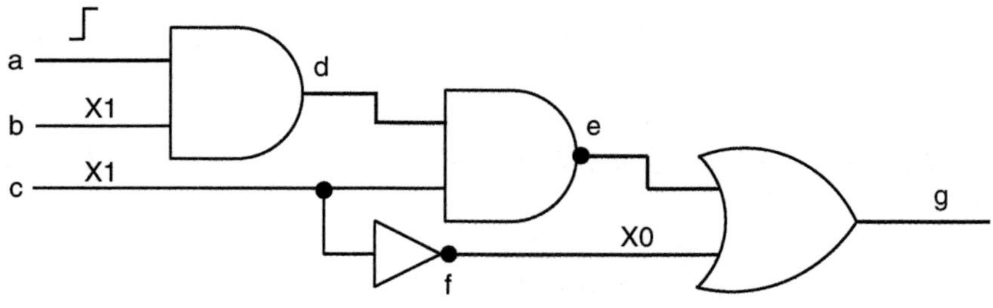

Figure 10.4: Example circuit for path a–d–e–g

Finally, the fault specific constraints are added. In contrast to the TDFM, the fault specific constraints can be considered as fixed assignments to variables and are divided into two parts. The transition must be launched at g_1 (Φ_{tran}) and the off-path inputs of P must be assigned according to the non-robust sensitization criterion as given in Table 10.1 (denoted by Φ_o).

$$\Phi_P = \Phi_{C_{NR}} \cdot \Phi_{tran} \cdot \Phi_o$$

If Φ_P is satisfiable, P is a non-robustly testable path and the test can be created directly from the calculated solution.

Example 23 *Consider the example circuit shown in Figure 10.4 with $P = (a, d, e, g)$ and $T = R$. The CNF of the circuit is as follows:*

$$\Phi_{C_1} = \Phi^d_{\mathrm{AND}_{t_1}} \cdot \Phi^e_{\mathrm{NAND}_{t_1}} \cdot \Phi^f_{\mathrm{NOT}_{t_1}} \cdot \Phi^g_{\mathrm{OR}_{t_1}}$$
$$\Phi_{C_2} = \Phi^d_{\mathrm{AND}_{t_2}} \cdot \Phi^e_{\mathrm{NAND}_{t_2}} \cdot \Phi^f_{\mathrm{NOT}_{t_2}} \cdot \Phi^g_{\mathrm{OR}_{t_2}}$$

Because no flip-flops are contained in this circuit, the equation $\Phi_{seq} = 1$ holds. The fault specific constraints for the rising transition are:

$$\Phi_{tran} = (\overline{a}_{t_1}) \cdot (a_{t_2}), \quad \Phi_o = (b_{t_2}) \cdot (c_{t_2}) \cdot (\overline{f}_{t_2})$$

A corresponding test given by the solution of the SAT solver could be:

$$v_1 = \{a_{t_1} = 0, b_{t_1} = 0, c_{t_1} = 0\}$$
$$v_2 = \{a_{t_2} = 1, b_{t_2} = 1, c_{t_2} = 1\}$$

10.2.2 Robust Test Generation

According to the robust sensitization criterion, static values have to be modeled. Therefore, Boolean values are not sufficient for robust test generation. Using only Boolean values, two discrete points of time t_1, t_2 are modeled, but no information about the transitions between t_1 and t_2 is given. The following example motivates the use of a multiple-valued logic.

Example 24 *Consider the AND gate in Figure 10.5a. If the robust sensitization criterion requires that the output is set to* $S0$*, it is not sufficient to set both output variables corresponding to the two time frames to* 0*. Then, a rising and a falling transition on the inputs would satisfy the condition, because the controlling value is assumed in* t_1, t_2 *on different inputs. However, if the inputs do not switch simultaneously, which cannot be guaranteed without timing information, a glitch could be produced on the output.*

This case can be excluded by explicitly modeling static values. This ensures that a static value on the output of a gate has its source in one or more static values on the inputs. This is shown at the AND gate in Figure 10.5b.

Static values can be handled using the multiple-valued logic

$$L_{6s} = \{S0, 00, 01, 10, 11, S1\}.$$

The name of the value determines the signal's behavior in t_1 and t_2. The first position gives the value of the connection in t_1, whereas the second position describes the value in t_2. The values $S0$ and $S1$ represent the static values. The truth table for an AND gate modeled in L_{6s} is presented in Table 10.2.

As described for the SAFM in Chapter 6, to apply a Boolean SAT solver to a problem formulated in multiple-valued logic, each value must be encoded using Boolean variables. This encoding is not unique. Among all possibilities a good encoding has to be found. Here, the encoding that turned out to be

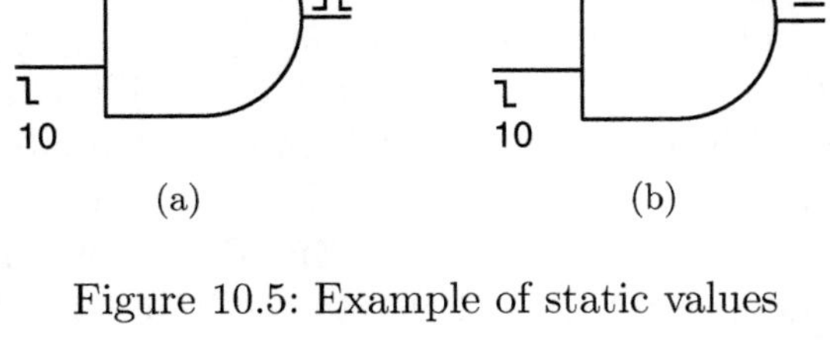

Figure 10.5: Example of static values

Table 10.2: Truth table for an AND gate using L_{6s}

AND	$S0$	00	01	10	11	$S1$
$S0$	$S0$	$S0$	$S0$	$S0$	$S0$	$S0$
00	$S0$	00	00	00	00	00
01	$S0$	00	01	00	01	01
10	$S0$	00	00	10	10	10
11	$S0$	00	01	10	11	11
$S1$	$S0$	00	01	10	11	$S1$

Table 10.3: Boolean encoding $\eta_{L_{6s}}$ for L_{6s}

Var	$S0$	00	01	10	11	$S1$
x_1	0	1	0	1	0	0
x_2	0	1	1	1	1	0
x_3	0	0	0	1	1	1

Table 10.4: CNF for an AND gate using $\eta_{L_{6s}}$

$$(\overline{x}_1^a + x_1^c + \overline{x}_2^c) \cdot (x_2^b + x_3^b + \overline{x}_2^c) \cdot (\overline{x}_1^a + x_2^a) \cdot$$
$$(\overline{x}_1^b + x_1^c + \overline{x}_2^c) \cdot (\overline{x}_2^a + \overline{x}_3^b + x_2^c) \cdot (\overline{x}_1^b + x_2^b) \cdot$$
$$(\overline{x}_3^a + \overline{x}_3^b + x_3^c) \cdot (\overline{x}_3^a + \overline{x}_2^b + x_2^c) \cdot (x_3^b + \overline{x}_3^c) \cdot$$
$$(x_1^a + x_1^b + \overline{x}_1^c) \cdot (\overline{x}_2^a + \overline{x}_2^b + x_2^c) \cdot (x_3^a + \overline{x}_3^c) \cdot$$
$$(x_2^a + x_3^a + \overline{x}_2^c) \cdot (x_2^a + x_2^b + \overline{x}_2^c) \cdot (\overline{x}_1^c + x_2^c)$$

the most effective one is chosen to illustrate the overall flow. The detailed explanation how to generate an efficient encoding for robust tests for the PDFM will be given in Section 10.3.

A logarithmic encoding is used. The minimal number of Boolean variables n needed to encode a value depends on the number of values of a multiple-valued logic L_m and is calculated as follows:

$$n = \lceil \log_2 |L_m| \rceil .$$

As a result, three variables are needed to encode each value of L_{6s}. The Boolean encoding $\eta_{L_{6s}}$ for L_{6s}, used in this book, is shown in Table 10.3. For example, the connection c has three variables x_1^c, x_2^c, x_3^c. Hence, an assignment $\{x_1^c = 0, x_2^c = 0, x_3^c = 1\}$ is interpreted as the value $S1$ of L_{6s}.

The resulting CNF for a 2-input AND gate with inputs a, b and output c using $\eta_{L_{6s}}$ is presented in Table 10.4. The CNF can be created using a truth table and a logic minimizer, e.g. ESPRESSO [89]. The SAT formulation of the circuit using L_{6s} is similar to the SAT formulation described in Section 10.2.1. However, instead of two variables, three variables are assigned to each connection and the circuit CNF is derived from the CNF of each gate using $\eta_{L_{6s}}$. The robust sensitization criterion is modeled by fixing the corresponding assignments. Note, that there is no need to build the CNF for the complete circuit. For a specific PDF, only the fanin cone of gates on the path has to be transformed into CNF using L_{6s}. If flip-flops are contained in this fanin cone, the fanin cone of these flip-flops has to be considered, too. By this, the sequential behavior is modeled adequately.

Let FF be the set of flip-flops contained in the fanin cone of the target path P. For all gates located in the fanin cone of at least one flip-flop in FF but not in the fanin cone of P, only the value during t_1 is relevant. Therefore, these gates can be modeled in Boolean logic as described in Chapter 3. Consequently, only one variable is needed. If a predecessor f of such a gate g is modeled in L_{6s}, only x_3^f is used for Φ_g. As Table 10.3 shows, $\eta_{L_{6s}}$ was chosen such that the assignment of x_3 always determines the value of the connection in t_1.

In contrast to the pure Boolean modeling of the circuit, using L_{6s} causes a serious overhead for the CNF size of the circuit. On the other hand, tests with a higher quality can be obtained and, as the experimental results in Section 10.5.3 below show, test generation for robust tests can be executed in reasonable time.

10.2.3 Industrial Application

In this section, the problem of PDF test generation in industrial practice is considered. Additional constraints that have to be handled in industrial circuits are introduced and structural techniques to reduce the size of the SAT instance are presented.

Additional Values

For industrial applications, more requirements have to be met for PDF test generation. As in the case of the SAFM (see Section 6.1), besides the Boolean values, two additional values have to be considered. The value Z describes the state of high impedance and occurs for example in modeling busses. Gates that are able to assume Z are named tri-state gates. If a connection has a fixed value which is not known, the value U (unknown) is assumed. The unknown value can for instance occur if some flip-flops cannot be controlled. Then, the output of the flip-flop always has the value U, i.e. it is fixed to U.

A test generation algorithm must be able to handle these additional values when it is applied in industrial practice. In Chapter 6, a four-valued logic $L_4 = \{0, 1, Z, U\}$ was presented. For PDF test generation, two time frames have to be considered; L_4 is not sufficient. Therefore, the Cartesian product of all values in L_4 is needed to represent all possible value combinations on a connection. For non-robust PDF test generation, this results in the 16-valued logic L_{16}:

$$L_{16} = \{00, 01, 10, 11, 0U, 1U, U0, U1, UU, \\ 0Z, 1Z, Z0, Z1, UZ, ZU, ZZ\}$$

As described in Section 10.2.2, additional static values are needed for robust PDF test generation. Therefore, the 19-valued logic L_{19s} is proposed:

$$L_{19s} = \{S0, 00, 01, 10, 11, S1, 0U, 1U, U0, U1, UU, \\ 0Z, 1Z, Z0, Z1, UZ, ZU, ZZ, SZ\}$$

The logic L_{19s} contains three additional static values: $S0, S1, SZ$. A static U value is meaningless, because the behavior of the signal is unknown.

In principle, L_{19s} can be used to model the circuit for robust PDF test generation. However, logics with fewer values are generally more compact in their CNF representation than logics with more values. The exclusive use of L_{19s} would result in excessively large SAT instances. This also holds for non-robust test generation using L_{16} exclusively. Fortunately, typically only a few connections in a circuit can assume all values contained in L_{19s} or in L_{16}. For example, there are only very few gates in a circuit that are able to assume the value Z.

Therefore, it is proposed to use not only one multiple-valued logic (e.g. L_{19s}) but a set of multiple-valued logics which are derived from L_{19s} (robust test generation) or L_{16} (non-robust test generation), respectively. These derived logics contain a smaller number of values, i.e. a subset of values. The idea behind this approach is that each gate is modeled using a logic that contains only those values which can be assumed by the input and output connections of the gate. This approach is an extension of the hybrid logic used for test pattern generation for stuck-at faults presented in Section 7.1. Using this approach, the size of the CNF is reduced.

Logic Classes

All gates that can always assume the same set of values are grouped into one logic class and are modeled in the same logic. Four different logic classes

Table 10.5: Derived logics of L_{19s} and L_{16}

L_{11s}	=	$\{S0, 00, 01, 10, 11, S1, 0U, 1U, U0, U1, UU\}$
L_{8s}	=	$\{S0, 00, 01, 10, 11, S1, 0U, 1U\}$
L_{6s}	=	$\{S0, 00, 01, 10, 11, S1\}$
L_9	=	$\{00, 01, 10, 11, 0U, 1U, U0, U1, UU\}$
L_6	=	$\{00, 01, 10, 11, 0U, 1U\}$
L_{4B}	=	$\{00, 01, 10, 11\}$

are identified for the classification of gates:

$$LC_Z, LC_{U1}, LC_{U2}, LC_B$$

In the following, the properties of each logic class are described as well as the dedicated logics that are used. Note, that for each logic class, two different logics can be used according to the desired quality of the test, i.e. non-robust or robust. The derived logics are presented in Table 10.5.

- LC_Z – A gate g belongs to LC_Z if all values of L_4 can be assumed in t_1, t_2. Obviously, only tri-state gates belong to this class. As described above, for robust test generation, L_{19s} is used, whereas for non-robust test generation L_{16} is applied.

- LC_{U1} – A gate g belongs to LC_{U1} if the values $0, 1, U$ can be assumed in t_1, t_2, but not Z. These gates are modeled using the derived logics L_{11s} (robust) and L_9 (non-robust).

- LC_{U2} – A gate g belongs to LC_{U2} if the values $0, 1$ can be assumed in t_1, t_2, whereas U can be assumed only in t_2. The corresponding logics are L_{8s} for robust test generation and L_6 for non-robust test generation.

- LC_B – A gate g belongs to LC_B if only $0, 1$ can be assumed in t_1, t_2. Then, L_{6s} (robust) and L_{4B} (non-robust) are applied. Note, that these gates are modeled as described in Sections 10.2.1 and 10.2.2.

A summary of the mapping between logic class and applied logic is given in Table 10.6. How to classify each gate is discussed in detail in Section 10.2.4.

Table 10.6: Mapping between logic class and applied logic

Logic class	Robust	Non-robust
LC_Z	L_{19s}	L_{16}
LC_{U1}	L_{11s}	L_9
LC_{U2}	L_{8s}	L_6
LC_B	L_{6s}	L_{4B}

Table 10.7: Boolean encoding $\eta_{L_{19s}}$ for L_{19s}

Var	$S0$	00	01	10	11	$S1$	0U	1U	U0	U1	UU	0Z	1Z	Z0	Z1	UZ	ZU	ZZ	SZ
x_1	0	1	0	1	0	0	1	1	1	0	1	0	1	0	1	1	0	0	1
x_2	0	1	1	1	1	0	0	0	1	0	0	1	0	0	0	1	0	0	1
x_3	0	0	0	1	1	1	0	1	1	1	1	1	1	1	0	1	0	0	0
x_4	1	1	1	1	1	1	1	1	0	0	0	0	0	0	0	1	1	0	0
x_5	0	0	0	0	0	0	0	0	0	0	0	1	1	1	1	1	1	1	1

Boolean Encoding

For the Boolean encoding of L_{19s}, five Boolean variables are needed to encode each value. The encoding[2] used, $\eta_{L_{19s}}$, is shown in Table 10.7. The encoding of the derived logics is implied by this encoding. For example, the encoding of the value $S1$ is $(\overline{x}_1\, \overline{x}_2\, x_3\, x_4\, \overline{x}_5)$. For L_{11s}, which needs only four variables, $S1$ is encoded by the first four variables, i.e. by $(\overline{x}_1\, \overline{x}_2\, x_3\, x_4)$, whereas for L_{6s} and L_{8s}, the value is encoded by $(\overline{x}_1\, \overline{x}_2\, x_3)$. For this reason, these encodings are said to be *compatible* with each other. The procedure for the encoding of L_{16} and its derived logics is similar.

In Table 10.8, the impact of the different logics on the size of the CNF for two example gate types (AND and busdriver) is presented. The truth table of a busdriver for one time frame is shown in Table 10.9. In column *Logic*, the logic used is listed, whereas in column *#Var*, the number of variables of the encoding is given. In the columns entitled *#Cls*, the number of clauses for an AND gate and a busdriver are shown. Columns named *#Lit* report the number of literals. The table entries are empty when the logic is not applicable for the gate type, for example, a Z is interpreted as U for an AND gate, while a busdriver can always assume the value Z so only L_{19s} and L_{16} apply.

[2]For a detailed discussion of how to determine an efficient encoding see Section 10.3 below. The encoding presented in this section corresponds to the encoding named η_{L6mee} in Section 10.3.

Table 10.8: Size of CNF for different logics

		AND		Bus driver	
Logic	#Var	#Cls	#Lit	#Cls	#Lit
L_{19s}	5	–	–	114	561
L_{11s}	4	30	97	–	–
L_{8s}	3	21	71	–	–
L_{6s}	3	15	40	–	–
L_{16}	4	–	–	86	426
L_{9}	4	20	50	–	–
L_{6}	3	14	35	–	–
L_{4B}	2	6	14	–	–

Table 10.9: Truth table of busdriver

Control		0	1	Z	U
Data	0	Z	0	U	U
	1	Z	1	U	U
	Z	Z	Z	Z	Z
	U	Z	U	U	U

The overhead of using a higher-valued logic is significant. Modeling as many gates as possible in a logic with fewer values is therefore desirable.

10.2.4 Structural Classification

In this section, the algorithm that determines the logic class of each gate is given. The classification is only executed once in a preprocessing step. Therefore, the overhead is negligible. The pseudo-code for the structural classification is given in Algorithm 5. Note, that this algorithm is an extension of Algorithm 2 presented in Section 7.1 used to configure hybrid logic for stuck-at test pattern generation. Algorithm 5 classifies gates with respect to the multiple-valued logics proposed in Section 10.2.3. Furthermore, it considers the propagation of unknown values into the final time frame.

To begin, the tri-state gates are identified. Typically, the number of these gates is small compared to the number of Boolean gates. Because all of them can always handle all values in t_1 and t_2, they are inserted into logic class LC_Z (line 4). All inputs that are fixed to an unknown state are also identified and inserted into logic class LC_{U1} (line 5) because they can assume the value U in t_1 and t_2 but not the value Z.

Algorithm 5 Structural classification in logic classes

```
LogicClass LC_Z, LC_U1, LC_U2, LC_B = ∅
GateList l = ∅
GateList f = ∅
LC_Z.insert(all_tri_state_gates())
LC_U1.insert(all_inputs_fixed_to_unknown())
l.push(all_tri_state_gates())
l.push(all_inputs_fixed_to_unknown())
mark_as_seen(l.all())
while !l.empty() do
   Gate g = l.pop_first_element()
   for all succ ∈ g.all_successors() do
      if not_seen(succ) then
         mark_as_seen(succ)
         l.push(succ)
         LC_U1.insert(succ)
         if is_FlipFlop(succ) then
            f.push(succ)
         end if
      end if
   end for
end while
while !f.empty() do
   Gate g = f.pop_first_element()
   for all succ ∈ g.all_successors() do
      if not_seen(succ) then
         mark_as_seen(succ)
         f.push(succ)
         LC_U2.insert(succ)
      end if
   end for
end while
for all gate ∈ all_gates() do
   if not_seen(gate) then
      LC_B.insert(gate)
   end if
end for
```

The next step is to determine the output cone of both, the tri-state gates and the fixed inputs. All gates which have been inserted into a logic class so far can be considered as sources of unknown values in the circuit. Note, that a Z-value is interpreted as U for a Boolean gate. Consequently, each gate in each output cone of these gates can itself assume an unknown state in t_1 and t_2. Therefore, they must be inserted into logic class LC_{U1}. This is done by the while-loop in lines 9–21.

If an unknown value reaches a flip-flop in the initial time frame, it is propagated again in the final time frame (due to launch-on-capture). Therefore, these flip-flops that are inserted into LC_{U1} are temporarily stored (lines 16–17). Once all elements of LC_{U1} are determined, the stored flip-flops are processed again by the while loop in lines 22–31, where the output cone of each flip-flop is determined. If a gate in an output cone is not in LC_{U1} or LC_Z, the value U can only be assumed in t_2 but not in t_1. For that reason, it is inserted into LC_{U2}. The remaining gates cannot assume a non-Boolean value in t_1 or t_2. Therefore, they are inserted into LC_B (lines 32–26).

One problem arising from the use of different logics in modeling a circuit is the handling of logic transitions. A logic transition occurs if at least one direct predecessor of gate g is modeled in a different logic than g. Due to the different Boolean encodings for the logics, inconsistencies would occur at g. Therefore, inputs of g modeled in a logic with fewer values must be converted to the higher-valued logic of g. Inconsistencies are avoided by additional constraints. The following example demonstrates the procedure.

Example 25 *Consider a busdriver b that is modeled in L_{19s}. The control input of b, named c, is modeled in L_{19s}, too. The corresponding variables are $x_1^b, x_2^b, x_3^b, x_4^b, x_5^b$ and $x_1^c, x_2^c, x_3^c, x_4^c, x_5^c$, respectively. For the data input of b, named d, L_{8s} is applied. The three corresponding variables are x_1^d, x_2^d, x_3^d. To obtain a consistent CNF, d is converted to L_{19s} and two additional variables x_4^d, x_5^d are assigned. Due to the compatible encoding of L_{19s} and L_{8s}, it is straightforward to restrict d to the values of L_{8s}. Table 10.7 shows that fixing x_4^d to 1 and fixing x_5^d to 0 is sufficient.*

The above structural analysis significantly reduces the complexity of SAT-based PDF test generation for industrial circuits.

10.3 Encoding Efficiency for Path Delay Faults

As shown in Chapter 6, the Boolean encoding of a multiple-valued logic is not unique. The Boolean encodings differ in the resulting sizes of the CNF representations as well as in their efficiency for SAT-based ATPG. For the four-valued logic applied for stuck-at test pattern generation, $4! = 24$ different encodings can be used. However, these encodings can be grouped in three different sets and the efficiency of each set can be experimentally evaluated.

For the robust PDF test generation, a six-valued logic L_{6s} at least is needed as shown in this chapter. For L_{6s}, $8!/2 = 20{,}160$ different Boolean encodings are possible.[3] The number of potential Boolean encodings increases with the number of values of the multiple-valued logic. For L_{8s}, there are $8! = 40{,}320$ Boolean encodings, whereas for L_{11s}, there are more than one billion.

Experimentally evaluating all possible Boolean encodings and selecting the most efficient one is therefore not feasible. Some pre-selection must be done to identify efficient encodings. To study the impact of the encodings, inefficient encodings have to be determined as well. Note, that preliminary experiments have shown that due to the small number of gates that have to be modeled in L_{19s}, the change of the Boolean encoding of L_{19s} has nearly no impact on the run time. Therefore, Boolean encodings for L_{19s} are not discussed.

Typically, but not necessarily, a larger SAT instance results in higher run times of the SAT solver. Moreover, in the field of SAT-based ATPG, the SAT solver has to cope with thousands of smaller instances. Although the complexity of building a SAT instance is of linear size, the overhead is not negligible in the overall run time (see the run time analysis in Section 7.2.1.) Therefore, a Boolean encoding with a compact CNF representation is likely to perform well whereas a Boolean encoding with a large CNF representation probably has a poor performance.

Each gate type has a different CNF representation and a preliminary evaluation (Section 6.1.3) has shown that one single Boolean encoding may produce a compact representation for one gate type, whereas for other gate types (e.g. a busdriver), it may be the contrary. Due to the fact that most gates in a circuit are Boolean gates and not tri-state gates, the following analysis is only based on the size of the CNF representation for Boolean gates. Primitive gates also have the advantage that the size of their representation is very similar for a specific encoding.

The CNF sizes of the Boolean encodings are analyzed below. The compactness of the Boolean representation of each encoding e is denoted as C^e and is defined as a tuple $(|cls|, |lits|)$ that contains the accumulated number

[3] Here, only logarithmic encodings are considered.

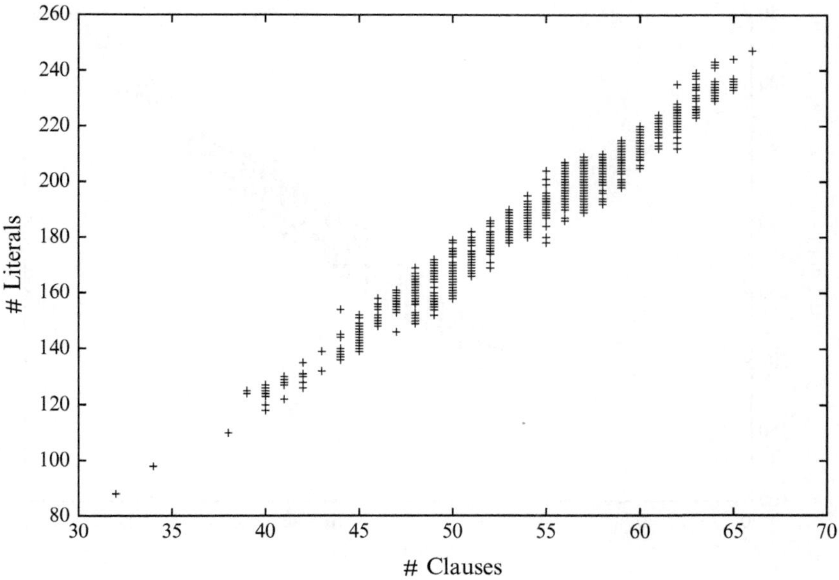

Figure 10.6: Distribution of the compactness values for Boolean encodings of L_{6s}

of clauses (cls) and the accumulated number of literals (lits) of the gate types AND and OR. The accumulation was done to obtain a good ratio of the compactness of both gate types.

The distribution of the compactness values of all possible Boolean encodings for L_{6s} and for L_{8s} are shown in Figures 10.6 and 10.7, respectively. The *Most Compact Encodings* (MCE) of L_{6s} have 32 clauses and 88 literals (accumulated for AND and OR), whereas the largest encodings have 67 clauses and 247 literals, which is more than two times the size of the MCEs, and concerning the number of literals nearly three times. The difference between the most compact and the largest encoding is even greater for L_{8s}. Here, the number of clauses in the largest encoding (97) is 2.6 times the most compact one (38) and the number of literals (448) is 3.8 times larger than the most compact one (118).

10.3.1 Compactness of Boolean Representation

Due to the very high number of possible encodings for L_{11s}, the range of the compactness values for the encodings of L_{11s} is determined with a simplified method. The compactness values are calculated only for those encodings of

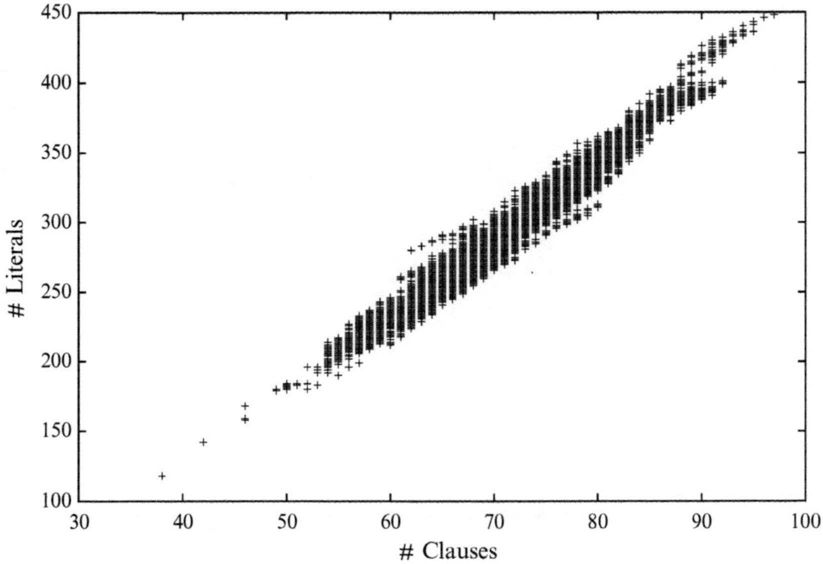

Figure 10.7: Distribution of the compactness values for Boolean encodings of L_{8s}

L_{11s} that are compatible with the MCEs of L_{6s} and L_{8s}. This is suitable since only compatible encodings can be used without additional overhead to model logic transitions (see Section 10.2.3). The distribution of the compactness values for these encodings of L_{11s} are shown in Figure 10.8. For this small subset of 64,512 encodings of L_{11s}, the number of clauses and the number of literals vary between 67 and 230 and 126 and 534, respectively.

An analysis of the MCEs of L_{6s} and L_{8s} shows that no compatible encoding of the MCEs of L_8 is among the MCEs of L_{6s}. Moreover, those encodings of L_{6s} that are compatible to the MCEs of L_{8s} have a larger size than the MCEs of L_{8s}. For example, consider the following compatible encodings η_e (L_{6s}) and η_f (L_{8s}). Whereas η_f with $C^f = (38, 118)$ is among the MCEs of L_{8s}, η_e with $C^e = (40, 118)$ is not among the MCEs of L_{6s} and is larger than η_f.

It can be concluded that the chosen Boolean encoding has – independent of the logic used – an enormous impact on the size of the SAT instance. The use of compatible encodings only, however, puts tight constraints on the use of Boolean encodings and prevents the joint use of the MCEs of each logic.

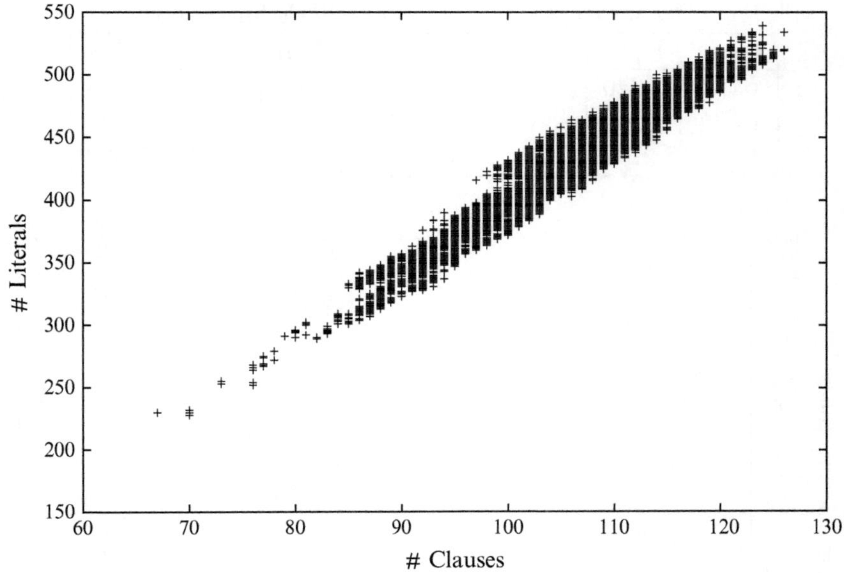

Figure 10.8: Distribution of the compactness values for Boolean encodings of L_{11s} (simplified)

10.3.2 Efficiency of Compact Encodings

The size of the SAT instance is only one indicator of the efficiency of a Boolean encoding. Therefore, the MCEs of each logic are investigated according to their run time for a single circuit. To avoid influences from other encodings, the circuit must be modeled by only one single logic, i.e. either L_{6s}, L_{8s} or L_{11s}. The ISCAS'85 circuit c6288, representing a 16-bit multiplier, was chosen to test the efficiency of the MCEs of each logic. All structural paths with a length of over 40 gates were identified and were set as targets (rising and falling) for robust test pattern generation. This results in 3,200 ATPG calls for each encoding.

The tests were carried out for each multiple-valued logic on an Intel Xeon (3 GHz, 32,768 MByte RAM, GNU/Linux). In each of the three runs, the circuit is modeled completely with L_{6s}, L_{8s} and L_{11s}, respectively. For each logic, a set containing the MCEs is identified and for each encoding in the set, robust PDF test pattern generation was performed. In Table 10.10, statistical data and the overall results of the runs are given. The first column gives the logic, whereas in the next column, the number of runs is shown. The third column presents the compactness values of the chosen encodings.

Table 10.10: Run times of MCEs for c6288

Logic	Runs	C^e	Min	Av.	Max
L_6	240	$(32, 88)$	68	113	285
L_8	144	(38–42, 118–142)	191	544	1647
L_{11}	192	$(67, 126)$	473	2,608	7,560

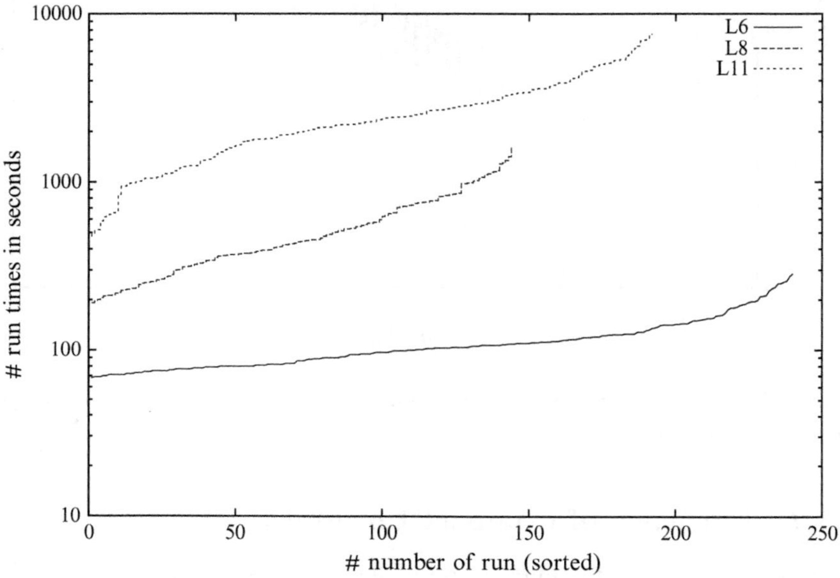

Figure 10.9: Run time distribution for c6288

The columns *Min*, *Av.* and *Max*, the smallest, the average and the highest run time, respectively, are given in CPU seconds.

In Figure 10.9, the run time distribution is shown for each logic on a logarithmic scale. The run times for each logic are sorted. The value on the x-axis defines the position in the sorted list and the value on the y-axis gives the run time in seconds. The upper curve denotes the run times of the MCEs of L_{11s}, whereas the middle curve and the lower curve give the run times of the MCEs of L_{8s} and L_{6s}, respectively.

For L_{6s}, even the MCEs differ significantly regarding their run time behavior. The highest run time is over four times the minimal one, although they have equal compactness values. The range is even higher for the higher-valued logics L_{8s} and L_{11s}. The highest run time for L_{8s} is eight times the minimal run time for L_{8s}, whereas for L_{11s}, the highest run time is nearly

16 times the minimal run time. While the curve of L_{6s} increases very slowly, the curves of L_{8s} and L_{11s} are steeper, suggesting that encodings of L_{8s} and L_{11s} have to be chosen more carefully. Note, that those encodings having the minimal run time for c6288 are denoted as *Most Efficient Encodings* (MEEs) in the following.

The application of the MCEs for robust PDF test generation shows first, that equal compactness values do not guarantee the same run time behavior, and second, that the impact on efficiency increases with a higher-valued logic.

10.3.3 Encoding Selection

In this section, Boolean encodings are created to determine the influence on the ATPG run time. The compactness values of each encoding can be found in Table 10.11. Note that if no logic is explicitly named in the following, an encoding refers to a set of compatible encodings for each logic rather than to a single encoding. Four different experiments are described below:

- Experiment 1 shows the behavior of two encodings of which one is likely to be very efficient, whereas the other is probably inefficient. A compact encoding η_{L6com} (MCE of L_{6s}) and a large encoding η_{L6lar} are chosen. Note, that the encoding of L_{6s} was first created and the compatible encodings are selected afterwards. Here, the most compact and the largest encodings, respectively, are selected among the compatible

Table 10.11: Compactness values of Boolean encodings

enc.	C^e (L_{6s})	C^e (L_{8s})	C^e (L_{11s})
η_{L6com}	$(32, 88)$	$(52, 184)$	$(115, 483)$
η_{L6lar}	$(64, 241)$	$(78, 347)$	$(161, 759)$
η_{L6med}	$(48, 162)$	$(60, 226)$	$(117, 465)$
η_{L11com}	$(32, 88)$	$(42, 142)$	$(67, 230)$
η_{L11lar}	$(32, 88)$	$(42, 142)$	$(113, 473)$
η_{L8com}	$(32, 88)$	$(42, 142)$	$(60, 190)$
η_{L8lar}	$(32, 88)$	$(70, 287)$	$(105, 413)$
η_{L6mee}	$(32, 88)$	$(42, 142)$	$(67, 230)$
η_{L11mee}	$(32, 88)$	$(42, 142)$	$(67, 230)$
η_{L8mee1}	$(40, 118)$	$(38, 118)$	$(56, 166)$
η_{L8mee2}	–	$(38, 118)$	$(60, 190)$

encodings. If not mentioned otherwise, this is the standard approach to choosing compatible encodings. Furthermore, an encoding η_{L6med} of medium size is selected.

- Experiment 2 shows the influence of the encoding selection for L_{11s} on the ATPG performance. For this purpose, a compact encoding η_{L11com} (MCE of L_{11s}) is created. Next, an encoding set η_{L11lar} is generated such that the encodings for L_{6s} and L_{8s} are equal to those of η_{L11com}, but instead of choosing an MCE of L_{11s}, the largest compatible encoding is selected.

- In Experiment 3, the influence of the encoding selection for L_{8s} is investigated. First, a compact encoding η_{L8com} is generated. Then, an encoding set η_{L8lar} is created containing the same encoding for L_{6s}, but with different encodings of L_{8s} and L_{11s}. Note, that possible differences in run time cannot clearly be attributed to the encoding of L_{8s}, because the encoding for L_{11s} also differs.

- In Experiment 4, the MEEs of each logic are evaluated for all circuits. This shows that to achieve good overall performance, it is not sufficient to use an encoding optimized for one logic only. Therefore, the encodings η_{L6mee} (MEE of L_{6s}), η_{L11mee} (MEE of L_{11s}) and η_{L8mee1} (MEE of L_{8s}) are created. As already stated in Section 10.3.1, all those encodings of L_{6s} that are compatible to the MCEs of L_{8s} have an even larger size than the MCEs of L_{8s}. Therefore, those parts of the circuit which may be modeled in L_{6s} are modeled in L_{8s} as well using η_{L8mee2}. Otherwise η_{L8mee1} and η_{L8mee2} are equal.

10.4 Incremental Approach

Generating robust tests for PDFs is desirable. Unfortunately, typically only few paths in a circuit are robustly testable. For those paths which are not robustly testable, a non-robust test is usually generated (if one exists). The approach considered so far in Sections 10.2.2 and 10.3, required two independent SAT instances for both types of tests. Both instances are optimized either for non-robust or robust test generation. A SAT instance built for non-robust test generation is not suitable for robust test generation, because static values are not modeled. On the other hand, a SAT instance built for robust test generation can generally be used for non-robust test generation but causes too much overhead for non-robust test generation.

The fact that robust as well as non-robust test generation is executed sequentially can be exploited by using incremental SAT. Therefore, a new incremental SAT formulation for the encoding of static values is presented.

The application of this incremental formulation is as follows. At first, a SAT instance Φ_{NR-p} for non-robust test generation is built for path p. If it is unsatisfiable, p is non-robustly untestable and, consequently, robustly untestable. If p is non-robustly testable, a SAT instance Φ_{R-p} for robust test generation is built. The SAT instance Φ_{R-p} is composed according to the following equation:

$$\Phi_{R-p} = \Phi_{NR-p} \cdot \Phi_{static}$$

The CNF Φ_{static} describes the static value justification of p. That means the separate modeling of static values in contrast to the logic modeling given by Φ_{NR-p}. Incrementally adding Φ_{static} to Φ_{NR-p} results in a SAT instance suitable for robust test generation and provides the following advantages.

- Build time – Instead of building a completely new SAT instance for robust tests, execution time is saved by reusing the existing SAT instance Φ_{NR-p}.
- Learned information – Conflict clauses created during non-robust test generation can be reused during robust test generation. As a result, large parts of the search space can be pruned.
- Structural information – According to the robust sensitization criterion, not all off-path inputs of p must be guaranteed to be static as defined in Table 10.1. Therefore, some parts of the circuit do not have to be included in Φ_{static}.

In the following, a description of how to derive Φ_{static} is given. At first, an additional variable x_S is assigned to each connection. This variable determines whether the signal on the connection is static ($x_S = 1$). If a static signal has to be forced on an off-path input g, x_S^g is fixed to 1. To justify this value, additional implications are added for each gate in $\mathcal{F}(g)$. For gate g with direct predecessors $h_1, \ldots, h_n$, these are as follows:

$$(x_S^g = 1 \wedge g = ncv) \quad \rightarrow \quad \prod_{i=1}^{n} x_S^{h_i} = 1$$

$$(x_S^g = 1 \wedge g = cv) \quad \rightarrow \quad \sum_{i=1}^{n} (x_S^{h_i} = 1 \cdot h_i = cv)$$

Table 10.12: Boolean encoding $\eta_{L_{16}}$ for L_{16}

Var	00	01	10	11	0U	1U	U0	U1	UU	0Z	1Z	Z0	Z1	UZ	ZU	ZZ
x_1	0	0	1	1	0	1	0	0	1	1	0	0	0	1	1	1
x_2	0	1	0	1	0	0	0	1	1	1	0	1	1	0	0	1
x_3	0	0	0	0	1	1	0	1	1	1	1	0	1	1	0	0
x_4	0	0	0	0	1	1	1	1	1	0	0	1	0	0	1	1

Table 10.13: CNF size of incremental SAT formulation

	Non-rob.		Inc.		Total	
Logic	#Cls	#Lit	#Cls	#Lit	#Cls	#Lit
AND						
L_9	20	50	18	58	38	108
L_6	14	35	16	52	30	87
L_{4B}	6	14	9	31	15	45
Bus driver						
L_{16}	86	426	25	81	111	507

If the non-controlling value (*ncv*) is on the output of g, all direct predecessors of g have to be statically non-controlling between both time frames, too. If the controlling value (*cv*) is on the output of g, at least one predecessor h_i has to be statically controlling. Thus, it is guaranteed, that a static value on an off-path input is justified. The additional implications are transformed into CNF according to the corresponding encoding of g. The Boolean encoding for L_{16} used in this book is shown in Table 10.12. The CNF sizes for Φ_{Static} and for $\eta_{L_{16}}$ are presented in Table 10.13. The CNF sizes of $\eta_{L_{16}}$ for an AND gate and for a busdriver are shown in column *Non-rob.* Column *Inc.* presents the CNF sizes for the incremental SAT formulation. In column *Total*, the accumulated CNF size for robust test generation is given. Compared to the CNF size for robust test generation presented in Table 10.8, the CNF for the incremental SAT formulation is only slightly larger. However, for L_6 and L_9, one more variable is needed compared to the corresponding logics L_{8s} and L_{11s}.

As mentioned above, not all gates have to be included in Φ_{static}. For example, if a rising transition occurs at an AND gate, the off-path inputs – and consequently their fanin cone – do not have to be considered. Let G_S^F be the set of gates on which static values must be guaranteed for the path delay fault F. Then, only those gates that are located in the fanin cone of at least one gate $g \in G_S^F$ are included in Φ_{static}. As a result, the size of the CNF for robust test generation can be further reduced.

Table 10.14: CNF for an AND gate using $\eta_{L_{4B}}$

$$(\overline{x}_1^a + \overline{x}_1^b + x_1^c) \cdot (x_1^a + \overline{x}_1^c) \cdot (x_1^b + \overline{x}_1^c) \cdot$$
$$(\overline{x}_2^a + \overline{x}_2^b + x_2^c) \cdot (x_2^a + \overline{x}_2^c) \cdot (x_2^b + \overline{x}_2^c)$$

Table 10.15: CNF description for static value justification for an AND gate using $\eta_{L_{4B}}$

$$(\overline{x}_2^a + x_1^b + \overline{x}_2^b + \overline{x}_S^c) \cdot (x_1^a + \overline{x}_2^a + \overline{x}_1^b + \overline{x}_S^c) \cdot (x_S^a + \overline{x}_2^b + \overline{x}_S^c) \cdot$$
$$(\overline{x}_1^a + \overline{x}_1^b + x_2^b + \overline{x}_S^c) \cdot (x_S^a + \overline{x}_1^b + \overline{x}_S^c) \cdot (\overline{x}_2^a + x_S^b + \overline{x}_S^c) \cdot$$
$$(\overline{x}_1^a + x_2^a + \overline{x}_2^b + \overline{x}_S^c) \cdot (\overline{x}_1^a + x_S^b + \overline{x}_S^c) \cdot (x_S^a + x_S^b + \overline{x}_S^c)$$

Example 26 *Consider an AND gate c with inputs a, b modeled in L_{4B}. The corresponding CNF is shown in Table 10.14. Given the additional variables x_S^a, x_S^b, x_S^c, the implications described above are presented in CNF in Table 10.15.*

This incremental approach requires some overhead when modeling a circuit, as the most compact encoding cannot always be chosen. Instead information about static values is added to the SAT instance for non-robust generation. However, there is the advantage that learned information can be reused.

10.5 Experimental Results

In this section, the experimental results for ATPG for delay faults are presented. The experimental results for the TDFM can be found in Section 10.5.1. Then the PDFM is considered. The effectiveness of the different encodings proposed in Section 10.3.3 is evaluated in Section 10.5.2. Then, the best encoding is used to evaluate robust and non-robust test generation. Finally, the incremental approach for PDF test generation is compared to the approach of using independent SAT instances. Statistical information about the industrial circuits used for the TDFM as well as for the PDFM can be found in Section 2.5. For all experiments, MiniSat [29] v1.14 was used as the SAT solver.

10.5.1 Transition Delay Faults

The techniques presented in the previous sections have been implemented in C++. Experimental results for this implementation on the ITC'99 benchmarks as well as on industrial circuits from NXP Semiconductors are

Table 10.16: Experimental results for the TDFM

Circ	#Targets	Untest.	Ab.	Time
b14	40,086	322	0	3:00 min
b15	38,094	1,632	0	4:31 min
b17	133,804	4,095	0	20:53 min
b18	459,360	35,846	2	2:10 h
b20	80,606	514	0	8:02 min
b21	82,060	567	0	8:37 min
b22	119,810	624	0	12:08 min
p44k	109,806	7,456	23	8:31 h
p49k	255,326	*Timeout*		
p80k	311,416	1,790	4	9:26 min
p88k	256,050	3,002	0	31:05 min
p99k	274,376	17,975	1	12:29 min
p177k	410,240	*Timeout*		
p462k	1,134,924	390,706	775	11:03 h
p565k	1,524,044	45,257	331	2:54 h
p1330k	2,464,440	97,811	32	12:25 h

provided in this section. The experiments were carried out on an Intel Xeon (3 GHz, 32,768 MByte RAM, GNU/Linux). Note, that a fault simulator is used for identifying other faults that are detected by a generated test as per normal industrial practice (fault dropping).

Table 10.16 presents the experimental results for ATPG for TDFs. In the first column, the name of the circuit is presented. Column *#Targets* gives the number of targets, i.e. TDFs, for which a test has to be generated. This number includes the faults dropped by the fault simulator. In column *Untest.*, the number of untestable faults is given, whereas in column *Ab.*, the number of faults, for which no test could be generated in the given time limit (20 CPU seconds) is shown. Column *Time* reports the overall ATPG run time in CPU minutes (*min*) or CPU hours (*h*). The overall timeout for a circuit was set to 20 h.

Although the TDFM is more complex than the SAFM, the results show that PASSAT is able to complete the test generation in most cases in reasonable time. Furthermore, only a few aborts were produced. Therefore, in addition to the SAFM, SAT techniques are well suited for the TDFM.

10.5.2 Encoding Efficiency for Path Delay Faults

In this section, the results of the four experiments described in Section 10.3.3 are presented. The experiments were carried out on an AMD64 4200+ (2.2 GHz, 2,048 MByte RAM, GNU/Linux). The program was implemented in C++. As benchmarks, ISCAS'85 circuits and industrial circuits were used.

Only paths with a length of over 40 gates are selected as test targets. The maximum number of test targets was set to 20,100. The paths are chosen randomly, but to avoid testing paths of only a small part of the circuit, at least one path starts at each input (if such a long path exists). The number of paths under test for each circuit are presented in Table 10.17 in column *PUT*. Furthermore, Table 10.17 provides information about how many elements are modeled in the respective multiple-valued logic. These numbers (in percent) can be found in the appropriately labeled columns.

In Table 10.18, the results of the selected encodings of Experiment 1 and Experiment 2 are shown. The results of the encodings of Experiment 3 and Experiment 4 are provided in Table 10.19. Time is measured in CPU minutes (*min*) and CPU hours (*h*), respectively. The timeout for each target

Table 10.17: Circuit statistics

Circuit	$\%L_{19s}$	$\%L_{11s}$	$\%L_{8s}$	$\%L_{6s}$	PUT
c1908	0	0	0	100	2,264
c2670	0	0	0	100	1,400
c3540	0	0	0	100	3,700
c5315	0	0	0	100	4,340
c6288	0	0	0	100	3,200
c7552	0	0	0	100	6,360
p44k	0	0	0	100	20,000
p49k	0	0	0	100	12,390
p57k	<0.1	0.2	25.7	74.0	20,000
p80k	0	0	0	100	20,078
p88k	0.6	7.3	21.4	70.7	20,052
p99k	0	1.5	5.6	92.9	20,026
p177k	0.8	27.4	54.5	17.3	20,028
p456k	0.1	25.5	73.5	0.9	20,070
p462k	0.2	13.6	34.7	51.5	20,006
p565k	6.5	7.3	59.0	27.2	20,088
p1330k	1.5	15.6	82.5	0.4	20,036

Table 10.18: Results for Experiments 1 and 2

Circuit	Exp. 1			Exp. 2	
	η_{L6com}	η_{L6lar}	η_{L6med}	η_{L11com}	η_{L11lar}
c1908	0:13 min	0:28 min	0:18 min	0:13 min	0:13 min
c2670	0:16 min	0:32 min	0:24 min	0:16 min	0:16 min
c3540	1:15 min	2:38 min	1:46 min	1:12 min	1:12 min
c5315	0:51 min	2:05 min	1:14 min	0:51 min	0:51 min
c6288	6:28 min	3:21 h	4:24 min	**2:16 min**	**2:16 min**
c7552	0:40 min	1:25 min	0:57 min	0:40 min	0:40 min
p44k	**2:07 h**	*Timeout*	2:29 h	2:25 h	2:25 h
p49k	3:13 h	*Timeout*	4:42 h	3:18 h	3:18 h
p57k	5:01 h	*Timeout*	**4:28 h**	9:00 h	8:59 h
p80k	**18:27 min**	17:18 h	23:20 min	45:47 min	45:47 min
p88k	**21:50 min**	*Timeout*	26:20 min	22:44 min	22:41 min
p99k	11:31 min	*Timeout*	13:20 min	11:11 min	**11:05 min**
p177k	1:24 h	10:30 h	1:27 h	1:11 h	1:18 h
p456k	1:10 h	19:08 h	1:15 h	53:17 min	1:04 h
p462k	43:04 min	1:53 h	43:44 min	39:45 min	43:59 min
p565k	8:12 min	11:56 min	8:59 min	7:19 min	7:18 min
p1330k	1:05 h	2:34 h	1:07 h	47:33 min	1:03 h

Table 10.19: Results for Experiments 3 and 4

Circuit	Exp. 3		Exp. 4			
	η_{L8com}	η_{L8lar}	η_{L6mee}	η_{L11mee}	η_{L8mee1}	η_{L8mee2}
c1908	0:12 min	0:12 min	**0:11 min**	0:12 min	0:15 min	0:14 min
c2670	0:15 min	0:15 min	**0:14 min**	**0:14 min**	0:18 min	0:18 min
c3540	1:11 min	1:11 min	**1:08 min**	**1:08 min**	1:25 min	1:24 min
c5315	0:47 min	0:47 min	**0:46 min**	0:47 min	0:59 min	0:59 min
c6288	2:17 min	2:17 min	**2:16 min**	2:51 min	4:48 min	4:58 min
c7552	0:37 min	0:37 min	**0:36 min**	0:37 min	0:47 min	0:46 min
p44k	2:15 h	2:15 h	2:16 h	*Timeout*	2:16 h	2:21 h
p49k	3:19 h	3:19 h	**3:07 h**	3:09 h	3:47 h	3:46 h
p57k	12:00 h	*Timeout*	9:59 h	*Timeout*	5:03 h	5:02 h
p80k	45:49 min	45:49 min	46:00 min	1:12 h	50:37 min	50:20 min
p88k	22:37 min	23:09 min	22:37 min	50:48 min	23:15 min	22:55 min
p99k	11:20 min	11:06 min	11:10 min	12:06 min	16:08 min	17:13 min
p177k	1:11 h	1:35 h	1:13 h	1:12 h	**1:05 h**	**1:05 h**
p456k	52:22 min	1:51 h	54:00 min	53:25 min	**46:01 min**	46:48 min
p462k	40:13 min	59:14 min	40:54 min	43:01 min	36:05 min	**36:00 min**
p565k	7:24 min	9:58 min	7:25 min	7:27 min	**6:54 min**	6:55 min
p1330k	44:01 min	1:11 h	45:38 min	46:55 min	**40:14 min**	40:42 min

was set to 20 CPU seconds, whereas the timeout for each ATPG run was 20 CPU hours. The minimum run time of all encodings of the four experiments is bolded for each circuit.

Experiment 1 shows that the influence of the Boolean encoding is significant. The run time for the large encoding η_{L6lar} significantly increases up to a factor of 56 (p80k) compared to η_{L6com} and up to a factor of 44 compared to η_{L6med}. In 5 out of 11 industrial circuits, η_{L6lar} even reaches the limit of 20 CPU hours. Therefore, η_{L6lar} is not feasible for industrial practice. Comparing η_{L6com} and η_{L6med}, the compact encoding is in most cases only slightly better than η_{L6med} and in two cases (c6288, p57k), it is worse.

In Experiment 2, the influence of the chosen encoding for L_{11s} is evaluated. In those circuits with no, or only a few, parts modeled in L_{11s}, the run times are the same or even slightly better using the large encoding η_{L11lar}. In circuits with higher percentage of L_{11s}, the maximum overhead is about 25% of the run time (p1330k).

In Experiment 3, the influence of the chosen encoding for L_{8s} is investigated. The results are similar to those of Experiment 2, but the impact on the run times is significantly greater. For p456k, where nearly three quarters of the circuit is modeled in L_{8s}, the run time increases by a factor of 2.9 and for p57k a timeout occurred.

In Experiment 4, the MEEs of each logic are investigated for all circuits. Among all encodings considered during the experiments, encoding η_{L6mee} is the best choice for all other ISCAS'85 circuits. But for the industrial circuits, η_{L6mee} provides the smallest run time only for p49k (completely in L_{6s}) and not any other circuit. Nonetheless, the results of η_{L6mee} are quite robust. In most cases, there is only a slight difference to the best result.[4]

Approach η_{L8mee2} has an advantage over η_{L8mee1} for the smaller circuits (with lesser percentage of L_{8s}), whereas η_{L8mee1} is better for the larger ones and therefore preferable. Encoding η_{L11mee} (MEE of L_{11s}) has only minimal performance gain for those circuits with a large portion of L_{11s} (e.g. p177k, p456k).

This experiment shows that the use of a Boolean encoding optimized for one multiple-valued logic only is not optimal because different logics are required to model a single circuit. Instead the choice should depend on the characteristics of the circuit, i.e. on the percentage of different logic classes in the circuit. For instance, the combination of $\eta_{L8mee1}/\eta_{L8mee2}$ and η_{L6mee},

[4]Due to the robustness, this encoding was used for robust PDF test generation presented in Section 10.2.

where η_{L6mee} is applied if the percentage of L_8 in the circuit is lower than 25% ($\eta_{L8mee1}/\eta_{L8mee2}$ otherwise), would provide the best result for eleven out of 17 circuits and is therefore more robust.

10.5.3 Robust and Non-robust Tests

In this section, experimental results for the PDFM are presented. The algorithm was implemented in C++ and integrated into the ATPG framework of NXP Semiconductors as a prototype. Experiments were performed on an AMD Opteron (2.8 GHz, 32,768 MByte RAM, GNU/Linux). Two different types of circuits are considered: ITC'99 benchmark circuits and industrial circuits provided by NXP Semiconductors.

For each circuit, 10,000 paths each with a length of more than 50 gates were randomly chosen (if they exist). Note, that these paths are chosen in such a way that no more than ten paths begin at the same input. If a test for a path could not be solved within twelve MiniSat restarts, the path is listed as aborted.

Statistical information about the structural classification for the industrial circuits is provided in Table 10.20. The first column gives the circuit's name. Besides Boolean gates, the industrial circuits contain primitives such as MUX gates. In column *#PIs* and *#POs* the number of primary inputs and primary outputs are given, respectively, whereas column #FFs presents the number of flip-flops. The last four columns show the results of the structural preprocessing of the circuit. In each of these columns, the percentage of gates contained in the identified logic class is presented.

Table 10.20: Information about the industrial circuits

Circ	#PIs	#POs	#FFs	$\%LC_Z$	$\%LC_{U1}$	$\%LC_{U2}$	$\%LC_B$
p44k	739	56	2,175	0	0	0	100
p57k	8	19	2,291	<0.1	0.2	25.7	74.0
p80k	152	75	3,878	0	0	0	100
p88k	403	183	4,302	0.6	7.3	21.4	70.7
p99k	167	82	5,747	0	1.5	5.6	92.9
p177k	768	1	10,507	0.8	27.4	54.5	17.3
p462k	1,815	604	29,205	0.2	13.6	34.7	51.5
p565k	996	201	32,409	6.5	7.3	59.0	27.2
p1330k	617	90	104,630	<0.1	9.0	15.0	75.9
p2787k	46,015	274	58,835	0.3	39.4	25.5	34.8
p3327k	4,093	274	148,184	3.5	16.1	80.2	0.2
p3852k	6,052	303	173,738	1.5	15.6	82.5	0.4

Table 10.21: Experimental results for PDF test generation

		Non-robust			Robust		
Circ	#Paths	#Nr	Ab.	Time	#Rob	Ab.	Time
b14	1,666	354	0	1:00 min	135	0	1:19 min
b15	3,696	346	0	2:02 min	74	0	3:09 min
b17	10,000	2,146	0	5:56 min	392	0	9:01 min
b18	10,000	2,259	0	6:23 min	510	0	8:24 min
p44k	10,000	6,800	0	34:26 min	1,784	0	49:59 min
p57k	10,000	2,428	4	43:50 min	1,532	116	2:23 h
p88k	10,000	5,793	0	6:54 min	1,606	0	8:39 min
p99k	10,000	5,262	0	5:09 min	620	0	5:57 min
p177k	10,000	960	73	6:26 h	117	99	2:38 h
p462k	10,000	1,537	0	11:17 min	336	0	12:08 min
p565k	10,000	4,001	0	8:42 min	211	1	6:23 min
p1330k	10,000	799	0	19:29 min	353	0	22:57 min
p2787k	10,000	830	0	44:31 min	110	0	40:23 min
p3327k	10,000	3,463	0	4:13 h	919	0	1:01 h
p3852k	10,000	2,133	7	7:02 h	645	1	3:38 h

The large circuits especially contain a large number of gates which are not in the Boolean logic class LC_B. For example, only 0.2% of all gates in p3327k can be modeled in Boolean logic. On the other hand, the number of gates in LC_Z is also very small.

Table 10.21 presents the experimental results for PDF test generation for the non-robust and robust configuration. The first column *Circ* gives the name of the circuit. In the upper part, results for circuits from the ITC'99 benchmark suite are presented, whereas in the lower part, results for the industrial circuits are shown. Column *#Paths* gives the number of paths for test generation, whereas in column *#Nr*, the number of non-robustly testable paths is shown. The results for each configuration are shown in the respective column. Columns denoted as *Ab.* contain the number of PDFs that could not be classified within the given restart limit.

Comparing non-robust and robust test generation, robust test generation is in most cases more time consuming than non-robust test generation – up to a factor of 3.3 for p57k. However, in those cases where non-robust test generation is particularly time consuming (p177k, p3327k, p3852), robust test generation is much faster – up to a factor of 4.2 (p3327k) – than non-robust test generation. This can be explained by the decreased fault coverage. Typically, untestable faults can be classified faster (see for comparison the run time analysis for stuck-at test pattern generation presented in Section 7.2.1).

10.5.4 Incremental Approach

In the following, two different approaches to generate the tests with highest quality are evaluated:

- Sequential – At first, non-robust test generation using L_{16} and its derived logics is performed. If the PDF is non-robustly testable, robust test generation is executed. Here, a completely new SAT instance is built using L_{19s} and its derived logics.
- Incremental – Similar to the sequential configuration. However, if the PDF is non-robustly testable, robust test generation is performed using the incremental SAT formulation.

For these experiments the same circuits and the same paths were considered as for the independent test generation of non-robust and robust tests described above. The results are shown in Table 10.22, where the columns

Table 10.22: Experimental results for high quality test generation

		Sequential			Incremental		
Circ	#Paths	#Rob	Ab.	Time	#Rob	Ab.	Time
b14	1,666	135	0	1:21 min	135	0	1:07 min
b15	3,696	74	0	2:23 min	74	0	2:07 min
b17	10,000	392	0	7:48 min	392	0	6:42 min
b18	10,000	510	0	8:25 min	510	0	7:05 min
p44k	10,000	1,784	0	1:20 h	1,784	0	1:34 h
p57k	10,000	1,530	4(114)	1:47 h	1,646	2(0)	59:13 min
p88k	10,000	1,606	0	12:17 min	1,606	0	8:40 min
p99k	10,000	620	0	8:07 min	620	0	6:10 min
p177k	10,000	102	73(51)	7:45 h	151	73(0)	6:32 h
p462k	10,000	336	0	12:44 min	336	0	12:28 min
p565k	10,000	211	0	12:32 min	211	0	12:09 min
p1330k	10,000	353	0	19:49 min	353	0	19:35 min
p2787k	10,000	110	0	47:37 min	110	0	44:13 min
p3327k	10,000	919	0	4:35 h	919	0	4:26 h
p3852k	10,000	639	7(1)	9:46 h	640	7(0)	7:41 h

are labeled in the same way as in the previous subsection. Additionally, for aborted faults the first number denotes the number of aborted non-robust tests, whereas the number in brackets gives the number of aborted robust tests for which a non-robust test could be generated. Columns labeled as *#Rob* give the number of robust tests that could be generated by each configuration. Columns denoted as *Time* provide the overall run time. Run time is given either in CPU minutes (*min*) or in CPU hours (*h*).

In comparison to the non-robust test generation, the sequential configuration needs more run time. This is due to the larger number of SAT instances that have to be solved. In terms of robust tests, the sequential configuration is slightly worse than the robust test generation for a three circuits – namely p57k, p177k and p3852k. The reason for the smaller number of robust tests is that non-robust test generation for a PDF was aborted and therefore, robust test generation was not executed for this particular PDF.

The incremental configuration is faster than the sequential configuration (except for p44k) and nearly as fast as the non-robust configuration. At the same time, due to the learned information and structural knowledge, even more robust tests were generated. Therefore, if non-robust and robust test generation are to be executed, the incremental configuration is the best choice.

Table 10.23 shows a direct comparison between robust and incremental configuration for generating robust tests. The SAT solver is called for the incremental configuration (column *Incr. (Rob.)*) only once and no learned clauses from a previous run, i.e. non-robust test generation, are available. In nearly all cases, the robust configuration (which is optimized for generating robust tests) is faster than the incremental configuration. However, p57k is an exception. Here, the incremental configuration is faster by a factor of 3. Additionally, no aborts were produced. In the case of p177k, the number of aborts decreases as well. However, at the same time, the run time increases. In contrast, the number of aborts increases for p3852k when using the incremental configuration.

If run time is important for robust test generation, the robust configuration is preferable. On the other hand, if robust coverage is important, the incremental configuration should be chosen. This configuration produces less aborts at the cost of longer run time. It also has the advantage that non-robust tests are automatically generated.

Table 10.23: Comparison between robust and incremental configuration

	Robust		Incr. (Rob.)	
Circ	Ab.	Time	Ab.	Time
b14	0	1:19 min	0	1:20 min
b15	0	3:09 min	0	3:27 min
b17	0	9:01 min	0	9:15 min
b18	0	8:24 min	0	9:05 min
p44k	0	49:59 min	0	1:02 h
p57k	116	2:23 h	0	49:59 min
p88k	0	8:39 min	0	9:00 min
p99k	0	5:57 min	0	6:04 min
p177k	99	2:38 h	40	3:14 h
p462k	0	12:08 min	0	13:21 min
p565k	1	6:23 min	0	8:47 min
p1330k	0	22:57 min	0	24:49 min
p2787k	0	40:23 min	0	39:40 min
p3327k	0	1:01 h	0	1:22 h
p3852k	1	3:38 h	14	4:00 h

10.6 Summary

In this chapter, SAT-based ATPG for delay faults has been presented in detail. Two delay fault models are relevant in industrial practice: the *Transition Delay Fault Model* and the *Path Delay Fault Model*.

At first, it has been shown how TDFs can be modeled in CNF by injecting stuck-at faults. The experimental results show that – although the TDFM is a more complex fault model – PASSAT was able to finish test generation in reasonable time providing good fault coverage.

Furthermore, a SAT-based approach for generating non-robust as well as robust tests for the PDFM has been presented. A six-valued logic has been proposed which is able to represent static values that are needed for guaranteeing robustness. For the industrial application, a 19-valued logic and a Boolean encoding have been presented. To reduce the large size of the resulting SAT instance using the 19-valued logic, a set of multiple-valued logics with a smaller number of values has been developed and a structural analysis has been applied. In this way, a large number of gates is modeled in a logic with fewer values. This reduces the size of the SAT instance significantly.

It also has been shown how incremental SAT can be used to speed up PDF test generation. A new incremental SAT formulation for static values has been presented that exploits the fact that non-robust and robust test generation are often executed sequentially. Using the incremental SAT formulation, information learned during non-robust test generation can be transferred to robust test generation and, by this, reduce the search space.

Finally, Boolean encodings for the multiple-valued logics for robust PDF test generation have been analyzed with respect to their efficiency. Due to the large number of possible Boolean encodings, experimentally evaluating all encodings is not practical. Therefore, a procedure to determine efficient encodings has been presented.

Chapter 11

Summary and Outlook

Guaranteeing the correctness of a manufactured circuit is an important task in industrial chip design flow. To ensure that the circuit functions correctly, test patterns are applied. The problem of generating test patterns for the postproduction test was considered in this book. Basic concepts and classical algorithms for *Automatic Test Pattern Generation* (ATPG) have been reviewed. Furthermore, the problem of Boolean Satisfiability (SAT) has been introduced and the basic algorithms and techniques have been explained.

Due to recent advances in SAT solving, modern SAT solvers have been applied successfully in various problem domains, e.g. verification and design debugging. How to generate test patterns for the *Stuck-At Fault Model* (SAFM) using a SAT instance such that the ATPG process benefits from the powerful Boolean reasoning embedded in modern SAT solvers has been considered. This basic approach and several improvements to create the SAT-based ATPG engine PASSAT have been described during the subsequent chapters.

SAT solvers benefit significantly from their use of learned information in terms of conflict clauses. Most of the learned information is circuit-specific and can be reused for other faults as well. Strategies to reuse dynamically learned information have been presented. These strategies improve the performance and reduce the number of faults that cannot be classified by SAT-based ATPG.

In industrial circuits, the ATPG algorithm has to deal with restrictions imposed by the circuit's environment as well as with tri-state elements. How to handle these additional constraints in a SAT-based approach using a four-valued logic and a Boolean encoding was explained in detail. The effectiveness of several Boolean encodings has been further studied. Typically, the larger a SAT instance is, the more run time a SAT solver needs for solving it (even though exceptions are possible). Therefore, procedures have been presented to reduce the size of the CNF formulas that have to be solved by the SAT solver. Besides the reduction of the CNF size, structural information has been incorporated into the decision heuristic resulting in a more robust search process.

We have further shown how a SAT-based ATPG algorithm can be integrated into an industrial ATPG framework. For this, the drawback of overspecified test pattern has been addressed and procedures to reduce the size of the test pattern have been presented. A combination of a classical structure-based ATPG engine and a SAT-based ATPG engine that exploits the advantages of both has been proposed. The combination of the two approaches results in a fast and highly robust ATPG framework.

Delay fault models have been considered as an extension of the developed SAT techniques. The *Transition Delay Fault Model* (TDFM) has been modeled in CNF by injecting a stuck-at fault. As a result, the efficient SAT techniques for the SAFM can be applied to the TDFM as well.

Further, SAT-based ATPG for the *Path Delay Fault Model* (PDFM) has been introduced. Multiple-valued logics have been applied to generate robust and non-robust tests for industrial circuits. Structural techniques have been proposed to reduce the size of the CNF formula. Moreover, an incremental formulation of static values exploits the fact that non-robust and robust test generation is performed sequentially. The influence of Boolean encodings has been further studied.

Several questions for future work can be identified. So far, today's most important fault models, i.e. SAFM, TDFM and PDFM, have been addressed using SAT-based ATPG. However, due to the steadily decreasing feature sizes and new manufacturing techniques, new fault effects have been observed that are not covered by these classical fault models. Hence, new fault models also consider parametric faults. Can the developed SAT-based approach be enhanced so that it is able to cope with these newly developed fault models?

Another problem is the test set size. With new fault models and the increasing size of circuits, the test set size, i.e. the number of test patterns, has become very large. Automatic test equipment has limited storage and the test set size is thus an important financial issue. Can a SAT-based ATPG approach reduce the test set size by creating more compact test patterns? Does a tight combination with test compression or test compaction techniques yield a decreased test set size?

In addition, moving to higher levels of abstraction is an important issue when considering today's design flows. Technical data coming from layout becomes more and more important to model fault effects that affect timing, cause bridging faults etc. Can a SAT-based tool be enhanced to incorporate more details while having an acceptable run time? How to achieve the trade-off between abstraction required by the SAT engine and accuracy of the model is an open and challenging problem.

At higher levels, the same questions are of interest. RT level ATPG has been considered for a long time and has some advantages. At the same time powerful reasoning engines that directly facilitate word level information during the proof process are being developed, often referred to as solvers for *Satisfiability Modulo Theory* (SMT). Do these solvers yield significant advantages for ATPG at higher levels of abstraction? Or are these the core engines to allow for more accurate models at lower levels of abstraction?

Another issue is the recent shift from single-core processors to multi-core processors. Common ATPG techniques were implemented single-threaded. Using only one core of a multi-core processor wastes CPU time. The development of multi-threaded ATPG techniques or SAT solvers, respectively, to exploit the multi-core architectures is of interest to reduce the increasing overall computation time.

This list of questions is not complete but offers some perspectives for future research in the domain of ATPG in general and SAT-based ATPG in particular.

Bibliography

1. M. Abramovici, P.R. Menon, and D.T. Miller. Critical path tracing - An alternative to fault simulation. In *Design Automation Conference (DAC)*, pages 214–220, 1983.

2. M.F. Ali, S. Safarpour, A. Veneris, M.S. Abadir, and R. Drechsler. Post-verification debugging of hierarchical designs. In *International Conference on Computer Aided Design (ICCAD)*, pages 871–876, 2005.

3. F.A. Aloul, I.L. Markov, and K.A. Sakallah. Shatter: Efficient symmetry-breaking for Boolean satisfiability. In *Design Automation Conference (DAC)*, pages 836–839, 2003.

4. D.B. Armstrong. A deductive method for simulating faults in logic circuits. *IEEE Transactions on Computers*, 21(5):464–471, 1972.

5. D. Bhattacharya, P. Agrawal, and V.D. Agrawal. Delay fault test generation for scan/hold circuits using Boolean expressions. In *Design Automation Conference (DAC)*, pages 159–164, 1992.

6. A. Biere. Adaptive restart strategies for conflict driven SAT solvers. In *International Conference on Theory and Applications of Satisfiability Testing (SAT)*, volume 4996 of *LNCS*, pages 28–33, 2008.

7. A. Biere, A. Cimatti, E. Clarke, and Y. Zhu. Symbolic model checking without BDDs. In *Tools and Algorithms for the Construction and Analysis of Systems (TACAS)*, volume 1579 of *LNCS*, pages 193–207, 1999.

8. Boolean Satisfiability Research Group at Princeton University. zChaff, 15 Dec. 2008. `http://www.princeton.edu/~chaff/zchaff.html`.

9. D. Brand. Redundancy and don't cares in logic synthesis. *IEEE Transactions on Computers*, 32(10):947–952, 1983.

10. M.A. Breuer and A.D. Friedman. *Diagnosis & Reliable Design of Digital Systems.* Computer Science Press, 1976.

11. F. Brglez, D. Bryan, and K. Kozminski. Combinational profiles of sequential benchmark circuits. In *IEEE International Symposium on Circuits and Systems (ISCAS)*, pages 1929–1934, 1989.

12. F. Brglez and H. Fujiwara. A neutral netlist of 10 combinational circuits and a target translator in fortran. In *IEEE International Symposium on Circuits and Systems (ISCAS)*, pages 663–698, 1985.

13. R.E. Bryant. Graph-based algorithms for Boolean function manipulation. *IEEE Transactions on Computers*, 35(8):677–691, 1986.

14. M.L. Bushnell and V.D. Agrawal. *Essentials of Electronic Testing for Digital, Memory and Mixed-Signal VLSI Circuits.* Springer, 2005.

15. C.-A.Wu, T.-H. Lin, C.-C. Lee, and C.-Y. Huang. QuteSAT: A robust circuit-based SAT solver for complex circuit structure. In *Design, Automation and Test in Europe (DATE)*, pages 1313–1318, 2007.

16. K. Chandrasekar and M.S. Hsiao. Integration of learning techniques into incremental satisfiability for efficient path-delay fault test generation. In *Design, Automation and Test in Europe (DATE)*, pages 1002–1007, 2005.

17. C. Chen and S.K. Gupta. A satisfiability-based test generator for path delay faults in combinational circuits. In *Design Automation Conference (DAC)*, pages 209–214, 1996.

18. K. Cheng and H. Chen. Classification and identification of nonrobust untestable path delay faults. *IEEE Transactions on Computer Aided Design of Circuits and Systems*, 15(8):845–853, 1996.

19. S.A. Cook. The complexity of theorem proving procedures. In *3. ACM Symposium on Theory of Computing*, pages 151–158, 1971.

20. F. Corno, M. Reorda, and G. Squillero. RT-level ITC 99 benchmarks and first ATPG results. In *IEEE Design & Test of Computers*, pages 44–53, 2000.

21. M. Davis, G. Logeman, and D. Loveland. A machine program for theorem proving. *Communication of the ACM*, 5:394–397, 1962.

22. M. Davis and H. Putnam. A computing procedure for quantification theory. *Journal of the ACM*, 7:506–521, 1960.

23. S. Disch and C. Scholl. Combinational equivalence checking using incremental SAT solving, output ordering, and resets. In *ASP Design Automation Conference (ASPDAC)*, pages 938–943, 2007.

24. R. Drechsler. BiTeS: A BDD based test pattern generator for strong robust path delay faults. In *European Design Automation Conference (EuroDAC)*, pages 322–327, 1994.

25. R. Drechsler. Using synthesis techniques in SAT solvers. In *ITG/GI/ GMM-Workshop "Methoden und Beschreibungssprachen zur Modellierung und Verifikation von Schaltungen und Systemen"*, pages 165–173, 2004.

26. R. Drechsler, S. Eggersglüß, G. Fey, A. Glowatz, F. Hapke, J. Schloeffel, and D. Tille. On acceleration of SAT-based ATPG for industrial designs. *IEEE Transactions on Computer Aided Design of Circuits and Systems*, 27(7):1329–1333, 2008.

27. N. Eén and A. Biere. Effective preprocessing in SAT through variable and clause elimination. In *International Conference on Theory and Applications of Satisfiability Testing (SAT)*, volume 3569 of *LNCS*, pages 61–75, 2005.

28. N. Eén and N. Sörensson. The MiniSat page, 15 Dec. 2008. `http://minisat.se/`.

29. N. Eén and N. Sörensson. An extensible SAT solver. In *International Conference on Theory and Applications of Satisfiability Testing (SAT)*, volume 2919 of *LNCS*, pages 502–518, 2003.

30. S. Eggersglüß and R. Drechsler. Improving test pattern compactness in SAT-based ATPG. In *Asian Test Symposium (ATS)*, pages 445–450, 2007.

31. S. Eggersglüß and R. Drechsler. On the influence of Boolean encodings in SAT-based ATPG for path delay faults. In *International Symposium on Multiple-Valued Logic (ISMVL)*, pages 94–99, 2008.

32. S. Eggersglüß, G. Fey, R. Drechsler, A. Glowatz, F. Hapke, and J. Schloeffel. Combining multi-valued logics in SAT-based ATPG for path delay faults. In *ACM & IEEE International Conference on Formal Methods and Models for Codesign (MEMOCODE)*, pages 181–187, 2007.

33. S. Eggersglüß, D. Tille, G. Fey, R. Drechsler, A. Glowatz, F. Hapke, and J. Schloeffel. Experimental studies on SAT-based ATPG for gate delay faults. In *International Symposium on Multiple-Valued Logic (ISMVL)*, 2007.

34. E.B. Eichelberger and T.W. Williams. A logic design structure for LSI testability. In *Design Automation Conference (DAC)*, pages 462–468, 1977.

35. G. Fey, S. Safarpour, A. Veneris, and R. Drechsler. On the relation between simulation-based and SAT-based diagnosis. In *Design, Automation and Test in Europe (DATE)*, pages 1139–1144, 2006.

36. G. Fey, J. Shi, and R. Drechsler. Efficiency of multi-valued encoding in SAT-based ATPG. In *International Symposium on Multiple-Valued Logic (ISMVL)*, pages 25–30, 2006.

37. G. Fey, T. Warode, and R. Drechsler. Using structural learning techniques in SAT-based ATPG. In *International Conference on VLSI Design (VLSID)*, pages 69–76, 2007.

38. K. Fink, F. Fuchs, and M.H. Schulz. Robust and nonrobust path delay fault simulation by parallel processing of patterns. *IEEE Transactions on Computers*, 41(12):1527–1536, 1992.

39. A.D. Friedman. Easily testable iterative systems. *IEEE Transactions on Computers*, 22:1061–1064, 1973.

40. Z. Fu, Y. Yu, and S. Malik. Considering circuit observability don't cares in CNF satisfiability. In *Design, Automation and Test in Europe (DATE)*, pages 1108–1113, 2005.

41. K. Fuchs, F. Fink, and M.H. Schulz. DYNAMITE: An efficient automatic test pattern generation system for path delay faults. *IEEE Transactions on Computer Aided Design of Circuits and Systems*, 10(10):1323–1335, 1991.

42. K. Fuchs, M. Pabst, and T. Rössel. RESIST: A recursive test pattern generation algorithm for path delay faults. In *European Design Automation Conference (EuroDAC)*, pages 316–321, 1994.

43. K. Fuchs, H.C. Wittmann, and K.J. Antreich. Fast test pattern generation for all path delay faults considering various tset classes. In *European Test Conference (ETC)*, pages 89–98, 1993.

44. H. Fujiwara and T. Shimono. On the acceleration of test generation algorithms. *IEEE Transactions on Computers*, 32:1137–1144, 1983.

45. M.K. Ganai, L. Zhang, P. Ashar, A. Gupta, and S. Malik. Combining strengths of circuit-based and CNF-based algorithms for a high-performance SAT solver. In *Design Automation Conference (DAC)*, pages 747–750, 2002.

46. M.A. Gharaybeh, M.L. Bushnell, and V.D. Agrawal. Classification and test generation for path-delay faults using single stuck-fault tests. In *International Test Conference (ITC)*, pages 139–147, 1995.

47. E. Gizdarski and H. Fujiwara. SPIRIT: A highly robust combinational test generation algorithm. *IEEE Transactions on Computer Aided Design of Circuits and Systems*, 21(12):1446–1458, 2002.

48. P. Goel. An implicit enumeration algorithm to generate tests for combinational logic. *IEEE Transactions on Computers*, 30:215–222, 1981.

49. E. Goldberg and Y. Novikov. BerkMin: A fast and robust SAT-solver. In *Design, Automation and Test in Europe (DATE)*, pages 142–149, 2002.

50. E. Goldberg and Y. Novikov. Verification of proofs of unsatisfiability for CNF formulas. In *Design, Automation and Test in Europe (DATE)*, pages 886–891, 2003.

51. L.H. Goldstein and E. L. Thigpen. SCOAP: Sandia controllability/observability analysis program. In *Design Automation Conference (DAC)*, pages 190–196, 1980.

52. D. Große and R. Drechsler. Acceleration of SAT-based iterative property checking. In *Correct Hardware Design and Verification Methods (CHARME)*, volume 3725 of *LNCS*, pages 349–353, 2005.

53. A. Gupta, A. Gupta, Z. Yang, and P. Ashar. Dynamic detection and removal of inactive clauses in SAT with application in image computation. In *Design Automation Conference (DAC)*, pages 536–541, 2001.

54. P. Gupta and M. Hsiao. ALAPTF: A new transition fault model and the ATPG algorithm. In *International Test Conference (ITC)*, pages 1053–1060, 2004.

55. J.N. Hooker. Solving the incremental satisfiability problem. *Journal of Logic Programming*, 15(1–2):177–186, 1993.

56. I.D. Huang and S.K. Gupta. Selection of paths for delay testing. In *Asian Test Symposium (ATS)*, pages 208–215, 2005.

57. F. Hutter, D. Babić, H.H. Hoos, and A.J. Hu. Boosting verification by automatic tuning of decision procedures. In *International Conference on Formal Methods in CAD (FMCAD)*, pages 27–34, 2007.

58. M.K. Iyer, G. Parthasarathy, and K.-T. Cheng. SATORI - A fast sequential SAT engine for circuits. In *International Conference on Computer Aided Design (ICCAD)*, pages 320–325, 2003.

59. N. Jha and S. Gupta. *Testing of Digital Systems.* Cambridge University Press, 2003.

60. H. Jin and F. Somenzi. CirCUs: A hybrid satisfiability solver. In *International Conference on Theory and Applications of Satisfiability Testing (SAT)*, volume 3542 of *LNCS*, pages 211–223, 2005.

61. J. Kim, J. Whittemore, J.P. Marques-Silva, and K.A. Sakallah. On applying incremental satisfiability to delay fault testing. In *Design, Automation and Test in Europe (DATE)*, pages 380–384, 2000.

62. K. L. Kodandapani and D. K. Pradhan. Undetectability of bridging faults and validity of stuck-at fault test sets. *IEEE Transactions on Computers*, C-29(1):55–59, 1980.

63. A. Krstić and K.-T. Cheng. *Delay Fault Testing for VLSI Circuits.* Kluwer, Boston, MA, 1998.

64. A. Kuehlmann, V. Paruthi, F. Krohm, and M.K. Ganai. Robust Boolean reasoning for equivalence checking and functional property verification. *IEEE Transactions on Computer Aided Design of Circuits and Systems*, 21(12):1377–1394, 2002.

65. W. Kunz. HANNIBAL: An efficient tool for logic verification based on recursive learning. In *International Conference on Computer Aided Design (ICCAD)*, pages 538–543, 1993.

66. W. Kunz and D.K. Pradhan. Recursive learning: A new implication technique for efficient solutions of CAD problems: Test, verification and optimization. *IEEE Transactions on Computer Aided Design of Circuits and Systems*, 13(9):1143–1158, 1994.

67. T. Larrabee. Test pattern generation using Boolean satisfiability. *IEEE Transactions on Computer Aided Design of Circuits and Systems*, 11:4–15, 1992.

68. H.K. Lee and D.S. Ha. HOPE: An efficient parallel fault simulator for synchronous sequential circuits. In *Design Automation Conference (DAC)*, pages 336–340, 1992.

69. H.K. Lee and D.S. Ha. Atalanta: An efficient ATPG for combinational circuits. Technical Report 12, Department of Electrical Engineering, Virginia Polytechnic Institute and State University, 1993.

70. M. Liffiton, M. Mneimneh, I. Lynce, Z. Andraus, J. Marques-Silva, and K. Sakallah. A branch and bound algorithm for extracting smallest minimal unsatisfiable subformulas. *Constraints: An International Journal*, to appear in 2009.

71. C.-J. Lin and S.M. Reddy. On delay fault testing in logic circuits. *IEEE Transactions on Computer Aided Design of Circuits and Systems*, 6(5):694–703, 1987.

72. X. Lin, K.-H. Tsai, C. Wang, M. Kassab, J. Rajski, T. Kobayashi, R. Klingenberg, Y. Sato, S. Hamada, and T. Aikyo. Timing-aware ATPG for high quality at-speed testing of small delay defects. In *Asian Test Symposium (ATS)*, 2006.

73. F. Lu. UCSB circuit SAT solver - A circuit SAT solver based upon signal correlation guided learning, 15 Dec. 2008. `http://cadlab.ece.ucsb.edu/downloads/CSAT.htm`.

74. F. Lu, L.-C. Wang, K.-T. Cheng, and R. Huang. A circuit SAT solver with signal correlation guided learning. In *Design, Automation and Test in Europe (DATE)*, pages 892–897, 2003.

75. F. Lu, L.-C. Wang, K.-T. Cheng, J. Moondanos, and Z. Hanna. A signal correlation guided atpg solver and its applications for solving difficult industrial cases. In *Design Automation Conference (DAC)*, pages 436–441, 2003.

76. J.P. Marques-Silva. The impact of branching heuristics in propositional satisfiability algorithms. In *9th Portuguese Conference on Artificial Intelligence (EPIA)*, pages 62–74, 1999.

77. J.P. Marques-Silva and K.A. Sakallah. Robust search algorithms for test pattern generation. In *International Symposium on Fault-Tolerant Computing (FTCS)*, pages 152–161, 1997.

78. J.P. Marques-Silva and K.A. Sakallah. Robust search algorithms for test pattern generation. Technical Report RT/02/97, Department of Informatics, Technical University of Lisbon, January 1997.

79. J.P. Marques-Silva and K.A. Sakallah. GRASP: A search algorithm for propositional satisfiability. *IEEE Transactions on Computers*, 48(5):506–521, 1999.

80. P.M. Maurer. Efficient event-driven simulation by exploiting the output observability of gate clusters. *IEEE Transactions on Computer Aided Design of Circuits and Systems*, 22(11):1471–1486, 2003.

81. K. Miyase and S. Kajihara. XID: Don't care identification of test patterns for combinational circuits. *IEEE Transactions on Computer Aided Design of Circuits and Systems*, 23(2):321–326, 2004.

82. M.W. Moskewicz, C.F. Madigan, Y. Zhao, L. Zhang, and S. Malik. Chaff: Engineering an efficient SAT solver. In *Design Automation Conference (DAC)*, pages 530–535, 2001.

83. T.M. Niermann and J.H. Patel. HITEC: A test generation package for sequential circuits. In *European Conference on Design Automation (EDAC)*, pages 214–218, 1991.

84. M. Panagiotis and D. Vroon. Efficient circuit to CNF conversion. In *International Conference on Theory and Applications of Satisfiability Testing (SAT)*, volume 4501 of *LNCS*, pages 4–9, 2007.

85. I. Pomeranz, S.M. Reddy, and P. Uppaluri. NEST: A nonenumerative test generation method for path delay faults in combinational circuits. *IEEE Transactions on Computer Aided Design of Circuits and Systems*, 14(12):1505–1515, 1995.

86. J.P. Roth. Diagnosis of automata failures: A calculus and a method. *IBM Journal Research and Development*, 10:278–281, 1966.

87. S. Safarpour, A. Veneris, R. Drechsler, and J. Lee. Managing don't cares in Boolean satisfiability. In *Design, Automation and Test in Europe (DATE)*, pages 260–265, 2004.

88. M. Schulz, E. Trischler, and T. Sarfert. SOCRATES: A highly efficient automatic test pattern generation system. *IEEE Transactions on Computer Aided Design of Circuits and Systems*, 7(1):126–137, 1988.

89. E. Sentovich, K. Singh, L. Lavagno, Ch. Moon, R. Murgai, A. Saldanha, H. Savoj, P. Stephan, R. Brayton, and A. Sangiovanni-Vincentelli. SIS: A system for sequential circuit synthesis. Technical Report, University of Berkeley, 1992.

90. O. Shacham and K. Yorav. On-the-fly resolve trace minimization. In *Design Automation Conference (DAC)*, pages 594–599, 2007.

91. J. Shi, G. Fey, R. Drechsler, A. Glowatz, F. Hapke, and J. Schlöffel. PASSAT: Efficient SAT-based test pattern generation for industrial circuits. In *IEEE Annual Symposium on VLSI (ISVLSI)*, pages 212–217, 2005.

92. O. Shtrichman. Pruning techniques for the SAT-based bounded model checking problem. In *Correct Hardware Design and Verification Methods (CHARME)*, volume 2144 of *LNCS*, pages 58–70, 2001.

93. A. Smith, A. Veneris, and A. Viglas. Design diagnosis using Boolean satisfiability. In *ASP Design Automation Conference (ASPDAC)*, pages 218–223, 2004.

94. G.L. Smith. Model for delay faults based upon paths. In *International Test Conference (ITC)*, pages 342–349, 1985.

95. P. Stephan, R.K. Brayton, and A.L. Sangiovanni-Vincentelli. Combinational test generation using satisfiability. *IEEE Transactions on Computer Aided Design of Circuits and Systems*, 15:1167–1176, 1996.

96. P. Tafertshofer and A. Ganz. SAT based ATPG using fast justification and propagation in the implication graph. In *International Conference on Computer Aided Design (ICCAD)*, pages 139–146, 1999.

97. P. Tafertshofer, A. Ganz, and K. Antreich. Igraine - An implication graph based engine for fast implication, justification, and propagation. *IEEE Transactions on Computer Aided Design of Circuits and Systems*, 19(8):907–927, 2000.

98. P. Tafertshofer, A. Ganz, and M. Henftling. A SAT-based implication engine for efficient ATPG, equivalence checking, and optimization of netlists. In *International Conference on Computer Aided Design (ICCAD)*, pages 648–655, 1997.

99. D. Tille and R. Drechsler. Incremental SAT instance generation for SAT-based ATPG. In *IEEE Workshop on Design and Diagnostics of Electronic Circuits and Systems (DDECS)*, pages 68–73, 2008.

100. D. Tille, G. Fey, and R. Drechsler. Instance generation for SAT-based ATPG. In *IEEE Workshop on Design and Diagnostics of Electronic Circuits and Systems (DDECS)*, pages 153–156, 2007.

101. S. Tragoudas and D. Karayiannis. A fast nonenumerative automatic test pattern generator for path delay faults. *IEEE Transactions on Computer Aided Design of Circuits and Systems*, 18(7):1050–1057, 1999.

102. G. Tseitin. On the complexity of derivation in propositional calculus. In *Studies in Constructive Mathematics and Mathematical Logic, Part 2*, pages 115–125, 1968 (Reprinted in: J. Siekmann, G. Wrightson (Ed.), Automation of Reasoning, Vol. 2, Springer, Berlin, 1983, pp. 466–483).

103. J.T. van der Linden. *Automatic Test Pattern Generation for Three-State Circuits.* Ph.D. thesis, Technical University Delft, The Netherlands, 1996.

104. J.A. Waicukauski, E. Lindbloom, B.K. Rosen, and V.S. Iyengar. Transition fault simulation. *IEEE Design & Test of Computers*, 4(2):32–38, 1987.

105. L.-C. Wang, J.-J. Liou, and K.-T. Cheng. Critical path selection for delay fault testing based upon a statistical timing model. *IEEE Transactions on Computer Aided Design of Circuits and Systems*, 23(11):1550–1565, 2004.

106. T. Warode. *Strukturelles Lernen in der erfüllbarkeits-basierten Testmustergenerierung (Structural learning for test pattern generation based on satisfiability).* Master's thesis, University of Bremen, Germany, 2006.

107. J. Whittemore, J. Kim, and K. Sakallah. SATIRE: A new incremental satisfiability engine. In *Design Automation Conference (DAC)*, pages 542–545, 2001.

108. M. J. Y. Williams and J. B. Angell. Enhancing testability of large-scale integrated circuits via test points and additional logic. *IEEE Transactions on Computers*, C-22(1):46–60, 1973.

109. K. Yang, K.-T. Cheng, and L.-C. Wang. Trangen: A SAT-based ATPG for path-oriented transition faults. In *ASP Design Automation Conference (ASPDAC)*, pages 92–97, 2004.

110. L. Zhang, C. F. Madigan, M. H. Moskewicz, and S. Malik. Efficient conflict driven learning in a Boolean satisfiability solver. In *International Conference on Computer Aided Design (ICCAD)*, pages 279–285, 2001.

111. L. Zhang and S. Malik. Validating SAT solvers using an independent resolution-based checker: Practical implementations and other applications. In *Design, Automation and Test in Europe (DATE)*, pages 880–885, 2003.

Index

GPSR Compliance

The European Union's (EU) General Product Safety Regulation (GPSR) is a set of rules that requires consumer products to be safe and our obligations to ensure this.

If you have any concerns about our products, you can contact us on ProductSafety@springernature.com

In case Publisher is established outside the EU, the EU authorized representative is:

Springer Nature Customer Service Center GmbH
Europaplatz 3
69115 Heidelberg, Germany

Batch number: 10583998

Printed by Printforce, the Netherlands